Specifying Minor Works

Efficient maintenance of a property requires a reliable assessment for defects or inadequacies and a systematic method for dealing with them. This book provides the information you'll need for both.

Throughout the specification process, all manner of issues face the surveyor, property manager or building engineer, from describing common defects to addressing energy efficiency and carbon emissions. In addition to these tasks the book also deals with:

- prioritising works
- the practicalities of specification
- building control.

Helping you navigate bureaucracy as well as tackling the practical challenges safely and effectively, this is a crucial guide for building engineers, surveyors, contractors and property managers.

Patrick Reddin is a director of Reddin & Company Limited, Chartered Building Surveyors and Corporate Building Engineers, London. He is a Past President of the Association of Building Engineers (ABE).

Specifying Minor Works

Edited by
Patrick Reddin

Routledge
Taylor & Francis Group

LONDON AND NEW YORK

First published 2014
by Routledge
2 Park Square, Milton Park, Abingdon, Oxon OX14 4RN

and by Routledge
711 Third Avenue, New York, NY 10017

Routledge is an imprint of the Taylor & Francis Group, an informa business

British Library Cataloguing in Publication Data
A catalogue record for this book is available from the British Library

Library of Congress Cataloging in Publication Data
Specifying minor works / edited by Patrick Reddin.
   p. cm.
   Includes bibliographical references and index.
   1. Buildings—Specifications. I. Reddin, Patrick.
   TH425.S66 2011
   692'.3—dc23
   2011036458

ISBN: 978–0–415–58351–0 (pbk)
ISBN: 978–0–203–12565–6 (ebk)

Typeset in Goudy
by Swales & Willis Ltd, Exeter, Devon

Contents

Notes on contributors

Arwel Griffith, BSc (Hons), FRICS, DipHI, PPBEng, FBEng, FIAS, P.Eng, FSPE, FNAEA, MTTS, MIoD, is Managing Partner of Robert Sterling Surveyors LLP. He is a Fellow of the Royal Institution of Chartered Surveyors (FRICS) having qualified for Fellowship by achievement in January 2001. He has been a member of the RICS since 1986. He is a former President of the Association of Building Engineers (ABE) and a former Council Member of the Society of Professional Engineers (SPE). In 2013, he received the prestigious Distinguished Service Award from the Chartered Association of Building Engineers. As 'Managing Partner' of Robert Sterling Surveyors LLP, his day-to-day surveying work involves the inspection, survey and valuation of residential and commercial properties and includes, in many instances, expert witness work, including appearances in court. He represented the Association of Building Engineers (ABE) on the Government's Housing Strategy Forum. In this regard he was heavily involved with the Home Information Packs Initiative. As part of this process, he assisted the Property Services National Training Organisation (PSNTO) with benchmarking standards for 'Home Inspectors' and was a member of the former Home Inspectors Certification Steering Group (HICSG). In the media, he is a regular contributor to both national and local television, and radio, and press and trade publications.

Patrick Reddin has worked in the surveying profession since 1963 and qualified as a Chartered Surveyor in 1971. He was President of the Association of Building Engineers for the year 2001-2, and Honorary Secretary from 2002 to 2009. His work involves alterations and repairs to existing buildings, both commercial and residential. A large part of his work arises out of diagnosis and rectification of building defects, including housing disrepair. A substantial proportion of this and his building design and specification work arises out of his Landlord Clients' responses to local authority Environmental Protection Act notices and to complaints from their tenants, sometimes resulting in litigation. He advises clients and other members of the building professions on defects and has reported on cases leading to litigation against professional advisers and contractors. He is included in the UK Register of Expert Witnesses, the Law Society Register of Experts and the Recommended List of Experts held by the Association of Personal Injury Lawyers. He is assessed by the University of Warwick as competent to perform an assessment under the Housing Health and Safety Rating System.

He is the author of *Dealing with Disrepair* published in the Arden's Housing Library series and of *State of Stock* published by the London Housing Federation in 1996. He is a member of the Editorial Board of *Landlord and Tenant Review*, published by Sweet &

Maxwell and was a member of the Law Commission Expert Working Group leading to the publication of the Issues Paper "Housing: Proportionate Dispute Resolution".

He has trained housing staff in Housing Disrepair and has been a visiting lecturer at Kings College London, training MSc students studying Environmental Health.

Rhys Taylor, PPBEng, FBEng, IEng, AMIStructE, MFPWS, has over 35 years experience in the areas of structural design, building pathology, structural movement/subsidence, party wall disputes, boundary disputes and general building surveying. Rhys also regularly acts as an Expert Witness for construction litigation cases and is a regular speaker at seminars on structural movement of buildings and The Party Wall etc. Act 1996. He has been proud to serve in several roles at regional and national level for the Association of Building Engineers (ABE) and in May 2008 he was inaugurated as President of the ABE. He is a Reader for the Queen's Anniversary Prize Awards for Education and for 5 years has been Chair of the panel of judges for the South Yorkshire and Humber LABC Building Excellence Awards. Rhys still lives and works in his home town of Rotherham in South Yorkshire, where he is a Director of Taylor Tuxford Associates, Consulting Building Engineers and Architectural Designers.

Michael Wadood, BSc (Hons), VPBEng, FBEng, FRICS, MCIOB, has over 20 years wealth of experience in the field of Building Control surveying, gained from his employment within local government and currently within private practice. During his period of employment with the London Borough of Brent, he started as a trainee and worked up to the position of a District Building Surveyor. While at the London Borough of Southwark he was a Principal Building Control Surveyor and currently he is a Director within MLM Building Control Ltd, a Corporate Approved Inspector. Michael has worked on a variety of projects which covers a number of market sectors, within London and the Home Countries. He also has experience in offering compliance consultancy work, fire safety design and general construction guidance. Michael strongly believes in professional development and was a part-time lecturer at the University of Westminster for many years. He has also given technical presentations to students at Anglia Ruskin University in Chelmsford as part of their module course. In addition he has presented papers on Building Regulations to outside bodies. Michael is the Vice President of the Association of Building Engineers and in May 2014 he will be inaugurated President of the Chartered Association of Building Engineers. He is also a member of the Chartered Institute of Building, where he sits on the Faculty of Building Control and Standards Board.

Preface

This is a co-operative book, written by four authors. There will inevitably be some variation in style between chapters, but we consider that this will increase interest rather than diminish it. The national building specifications are now available online for composition of a domestic specification, so the need to subscribe to an annual service is no longer a restriction. We aim therefore to provide an understanding of what needs to be considered when specifying works rather than the details of the specification clauses.

This book's predecessor was published in three editions under the authorship of Mr Leslie Gardiner. When the first edition was published in 1983 it was a pioneering work, setting out for the first time a method of specifying building works, which was accessible to and useable by the professional, the lay-person and the contractor. It provided an ordered basis for pricing works. The 2nd edition, published in 1986, updated and improved on the original work and extended the contractual sections to include the then newly introduced intermediate form of contract. The response to the book and its value in promoting method in the practice of specifying minor building works resulted in a 3rd edition in 1990.

This edition of the book is intended for:

- Non-technical workers in property management
- House owner/occupiers
- Property professionals
- Property maintenance workers
- Contractors.

House owners, occupiers, be they tenants or owners, managers of property, maintenance personnel, contractors and individual operatives are all in the front line in the battle to carry out effective works to buildings which both represent value for money and add value to those properties. They all in their various roles have to confront defects and inadequacies in existing buildings. Between them they have to carry out repairs and need to be equipped to adequately assess and respond to problems. They need to be armed not just with a means of diagnosing defects and inadequacies but also assessing the seriousness and possible consequences.

The authors are all members of the Association of Building Engineers. The editor and author of some chapters is Patrick Reddin; the other authors are: Rhys Taylor, Arwel Griffith, and Michael Wadood.

Note

From 1 January 2014 the Association of Building Engineers has become the Chartered Association of Building Engineers.

1 Statutory control

This chapter will review some of the aspects associated with 'statutory control', which in essence is the legislation that covers the construction of building work within the UK. When dwellings are being constructed, extended or altered it is important to know that consent from the various statutory authorities may need to be obtained. Such consent will be discussed further within this chapter, where we shall expand on the various processes that are required prior to commencement of work.

We will primarily feature 'planning consent' from the planning authority and 'Building Regulations consent' from a Building Control Body. We will also cover aspects of environmental health, highways and water authority consent. Guidance about the Party Wall, etc. Act 1996 is given in Chapter 4.

What is planning?

So what exactly is 'planning consent' I hear you ask and 'why do we need it?' In essence, 'planning legislation' exists to protect amenities and the environment for the public interest and not the individual person's preference. The legislation is implemented by the local authorities to ensure that the type of development proposed is appropriate for the area concerned, while ensuring that the character and the amenities of the area are not adversely affected by the proposals or changes in the use of existing buildings or land.

The planning authority prevents buildings and new uses that would be harmful to the public interest. Without it, homeowners could build extensions that block out all of their neighbours' light for example, or new buildings could, for instance, cause dangerous traffic conditions by blocking the 'line of sight' along a road.

Planning has a positive effect on the local environment, coordinating the development of homes and places of work. Planning ensures that these are accessible and built in the right places in the right way.

What are the Building Regulations (otherwise known as Building Control)?

So if that is 'planning', what is the purpose of Building Regulations? Some consider Building Regulations as rules that provide a detailed system of quality control at all stages of building work. Historically, arising as a result of poor housing conditions, the regulations have been regarded as a cornerstone in maintaining the general standards of public health. There is case law, which refers to previous judgments from the courts, to support this statement from *Anns* vs. *Merton London Borough Council* in 1978, where the judge stated:

the purpose of the Regulations is to provide for the health and safety of owners and occupiers of buildings, including dwelling houses, by setting standards to be complied with in construction and by enabling local authorities to supervise and control the operations of builders.

This view was later incorporated within the Building Act 1984 where Section 1 stated that the Building Regulations are minimum standards lay down by Parliament to secure the 'health and safety, welfare and convenience of persons in or about buildings and of others who may be affected by buildings or matters connected with buildings'. This section goes on to state that the Building Regulations are also there 'furthering the conservation of fuel and power' and 'preventing waste, undue consumption, misuse or contamination of water'.

Building Regulations are created by the Secretary of State as a vehicle to enable compliance with the Building Act 1984. Within the Act, Schedule 1 lists all of the Building Regulations and in England and Wales these are supported by guidance documents called 'Approved Documents'. In Scotland these are known as 'Technical Handbooks' and within Northern Ireland they are known as 'Technical Booklets'.

No matter where you are within the UK, almost every new building or structural change to an existing building will require Building Regulation approval. Anyone wishing to construct or alter a building must apply to a Building Control Body – either the local authority or an Approved Inspector – for Building Regulations approval.

What are Building Control Bodies?

You are probably thinking, 'What's an Approved Inspector?' Like the local authority, which employs Building Control surveyors, Approved Inspectors are companies within the private sector which are registered, insured and licensed to act by the Construction Industry Council and which also employ Building Control surveyors. Details can be found on the Construction Industry Council's website www.cic.org.uk. Approved Inspectors undertake the same role as local authority Building Control surveyors and will make an assessment of the plans and carry out site inspections to ensure the construction scheme complies with the Building Regulations.

Apart from one working within the private sector and the other within the public sector, there are other differences between the two, the main one being that an Approved Inspector is not confined to operating within a defined geographical area. An Approved Inspector can work anywhere within England and Wales but, currently, they are not registered to work in Scotland or Northern Ireland. They have the ability to work overseas, offering a 'compliance assessment' of a scheme which may be designed using UK standards/legislation and where the client seeks reassurance that the design team's approach would be acceptable if the same scheme was built within the UK. It is known that some Approved Inspectors offer other services or are part of a multi-discipline organisation that undertakes additional tasks like Fire Risk Assessments, Fire Safety Surveys and access consultancy audits, for example.

On their side, the Local Authority Building Control Service (LABC) has introduced a 'Partnership Arrangement Scheme' where an architect/agent can approach their local authority Building Control department to assess a scheme under the Building Regulations, even though it may not be within their borough. The partnership arrangement comes into play when the work is inspected by the local authority where the work is being carried out. Local authorities also offer other services which can be viewed on the LABC website.

What about enforcement?

'What will happen if construction work is undertaken without consent?' Unauthorised work can result in harsh penalties against both the contractor and the owner of the property. This could take the form of a fine or worse – removal of the offending work.

The tell-tale signs which can give it all away include: skips full of building waste outside a property, building materials on the road or pavement, scaffolding outside the house and potential complaints from aggrieved neighbours to the local authority. It only takes a phone call to the local authority Building Control surveyor or planning officer. Alternatively, either one of them could simply drive past the site and notice that unauthorised works are being undertaken.

Building Control contravention

If unauthorised works are discovered, the following actions may take place. Under the Building Act 1984, Section 35, local authority Building Control surveyors can take legal action through a magistrates' court against the contractor for not notifying them of any one or all of the offenses – from the commencement of work, undertaking foundation work, to covering over newly installed drainage, covering the oversite, and completing any building work which results in occupation of the building. Each offence, if proved, carries a fine. There is a further daily penalty if the contravention remains or the default continues after conviction. This course of action has a shelve life of 6 months.

Action against the owner could take the form of a notice issued by the magistrates' court under Section 36 of the Building Act 1984. This route of action gives the owners 28 days to either remove or alter the offending work. Failure to comply would result in the local authority undertaking the work in default, charging the owner in order to recover its costs. The notice needs to be issued to the owner within 12 months of the completion of the work. Where works have been carried out following plans which has been approved by the local authority, this route of action cannot be taken – even if the plans showed a contravention against the Building Regulations.

This only refers to local authority Building Control and not to Approved Inspectors. This is because within Section 91(2) of the Building Act 1984 it states that 'It is the function of local authorities to enforce building regulations in their areas'.

When Approved Inspectors consider that the proposed building work appears not to comply with the requirements of the Building Regulations they may decide not to issue their final certificate and are within their rights to cancel their initial notice and notify the local authority where the work is being undertaken. As a result, the function of Building Control will automatically transfer to the local authority, which will use its enforcement powers to require the work to be altered or removed.

Dangerous structures

A building can create a dangerous situation in a number of ways. A building can suffer from an 'ageing process' and lack of maintenance, which could result in aspects of the structure falling into disrepair and dilapidation. An example of this might be when roof tiles are missing from a pitched roof and over a period of time the timber rafters become exposed to the elements, leading to wet rot and, as a result, lose their structural capacity to bear a load, resulting in the collapse of the roof structure. A building can also become dangerous as a direct result of a fire, or storm damage, as well as a vehicle colliding with the building or an explosion within the building, caused by a gas leak for example.

Dangerous structures can also be created by those who undertake construction work who are unskilled in the task that they are doing, who are not being overseen by a competent person, or who are following design specifications, details or drawings that are incorrect or inappropriate to the scheme. No one wants to build a dangerous structure, but sometimes the lack of knowledge and experience of those who undertake construction projects results in potentially dangerous structures being created – if it wasn't for Building Control surveyors who regularly inspect building projects and offer guidance and sometimes strongly worded advice, I'm sure the death and injury rates within the construction industry, as well as those who live within the buildings created or altered, would be much higher.

Internal structural works like those mentioned below tend to carry the highest risk of creating a dangerous structure. This usually occurs when the work is not undertaken correctly or the guidance from professional people is not followed.

Chimney breast removal

Not all chimney breasts on a party wall (refer to Chapter 4 for work undertaken to a party wall) can be removed. In some houses it is possible that the projection of the breast from the wall on the ground floor is different to the breast on the first floor. Some houses tend to have a projection of 100 mm on the first floor and 250 mm on the ground floor. Where this condition arises, the flues of the chimney are within the party wall and if you were to remove your chimney breast, it is possible that you will expose and damage your neighbour's flue. This becomes a really vital issue if they are using their chimney for a gas fire, since the fumes from the appliance could enter into the neighbouring property. In addition, the removal of the breast might also reduce the fire resistance between the two properties. For more guidance you should refer to Approved Document J. If you did want to remove a chimney breast then one on an external wall rather than a party wall would be more easily accomplished. In addition, those where the projection is greater than 250 mm on the ground floor and similar size on the first floor would be feasible, subject to a full survey being undertaken. In these circumstances it is important to remove the hearth when the ground floor breast is being removed and make good the void in the floor created by the removed hearth, as well as support the remaining stack within the loft space by using both brackets and a concrete lintel or a steel beam supported on the external walls.

With regards to the chimney stack, you should ensure that you do not remove the brickwork above the 'purlin line' – this is the timber beam which runs at 90° to the timber roof rafters. When removing a chimney breast on an external wall, you should always ensure that you have the same amount of brick below the roof line as you can see above the roof. This is to maintain the loading back to the wall via the corbels. These two factors ensure stability of the stack. Chimney breast removal is classified as 'building work' under the Building Regulations and, as such, Building Control consent is required for this type of work.

Through-room alterations

A number of homeowners require enlarged floor areas and sometimes seek to remove the wall between the lounge and dining room. This wall tends to be at 90° to the party wall and also tends to be loadbearing, since the wall carries the support of the roof and upper floors. Before removing the wall, consideration should be given to ensure the capability of the retained wall to maintain its structural stability and transmit the weight of the roof and upper floor to the foundations. The opening created in the wall needs to be supported by a structural beam on a good padstone; this is either a few courses of Class B Engineering bricks, or a concrete cube which has been worked out for its size. The beam needs to be of a size and weight that has been calculated by a qualified building engineer, structural engineer or a building surveyor. Depending on the result of the calculations and size of the remaining pier, you may be required to brick up one of the doors to add structural stability or build a pier to support one end of the new beam. Failure to gain competent guidance could create a serious structural fault which in turn could evolve into a dangerous structure.

Internal walls should always be assessed before any work to them is undertaken, since they offer overall stability to other parts of the structure; this is particularly important where terraced houses are involved. Like chimney breasts, these works are classified as 'building works' and require Building Control consent.

Planning contravention

With regards to planning, all planning authorities have enforcement officers who seek out unauthorised work and take action in accordance with their set procedures. It should be noted that their available actions for dealing with breaches of planning control are set out in the Planning and Compensation Act 1991, which amended Part VII of the Town and Country Planning Act 1990.

The first stage is for the planning authority to serve a 'Requisition for Information'. This is a statutory notice requiring the recipient to complete and return a questionnaire within 14 days. This provides the planning authority with details of the owner or, if warranted, the planning authority is able to undertake searches with the Land Registry. In exceptional circumstances a formal Planning Contravention Notice may be served to ascertain the required details. Once served, the information required by the Notice must be supplied within 21 days.

If 10 working days has lapsed since the issue of either the Requisition for Information Notice or Planning Contravention Notice, the planning authority can then serve an Enforcement Notice – failure to comply with this requirement may result in prosecution. The Enforcement Notice will state what needs to be undertaken by the owner and the time period in which compliance must be met.

If the notice is ignored, formal action may result in court orders to either demolition or clear the site. In this event the local authority can seek to recover their costs from the person or place a charge on the land.

Potential contraventions could include:

- Building work, material change of use without planning consent.
- Not following the approved plans which had planning consent and building something different.
- Not fully complying with any condition or limitations which are part of a planning consent.
- Demolition of a building within a conservation area without consent.
- Undertaking any work to a listed building without consent.
- Removal of protected trees or hedgerows without consent.
- Installation of a display advertisement without consent.

Is there an appeal procedure?

It should be noted that within the UK legal system there is also the right of appeal. You can appeal against a planning decision to the Planning Inspectorate prior to an Enforcement Notice being issued or you can appeal to the court against a Building Regulation Notice, or obtain from a suitably qualified person a written report concerning work to which a notice relates.

History of 'planning' to date

The concept of planning consent has been around since the nineteenth century, when there was a dramatic increase in population and the focus of this growth in towns led to public health problems. This demanded a new role for central government and, as a result, it was necessary to interfere with market forces and private property rights in the interests of social well-being. In 1848 the government passed the Public Health Act and created the Central Board of Health, which also considered planning issues. In the years that followed, this Act was amended several times to reflect the needs of society. Finally, in 1909, the government introduced the Housing, Town Planning Act and since then the Act has been amended and the provisions within it have grown.

Over the years there has been much debate about whether the planning system should be used to prevent any changes to local environments, while others think that planning controls are an unnecessary interference with individuals' rights and visions.

Central government in England and Wales has set out the national planning policy (Scotland and Northern Ireland's planning policy is created by their regional government), but it has given the responsibility for planning consent to the local authorities, which are designated as the 'planning authority'. These authorities then create a 'planning policy' that provides a guide which controls and shapes the physical development of redevelopment in a particular area. This policy has been decided in conjunction with the local community, and it is constantly developing to take into account changing demands for housing, factories, offices and shops, and changes in technology and the way we live.

Planning has played an integral part in the development of the country's social and communal well-being. Over the years it has evolved and changed and currently has a positive effect on the local environment through the co-ordination of the development of homes and places of work, and helps to ensure that they are accessible and built in the right places in the right way.

The recent Planning Act 2008 has reformed the planning application process in England and Wales. The Act introduced a new streamlined system for decisions on applications in England and Wales. In Scotland the new Planning etc. (Scotland) Act 2006 is the central part of the most fundamental and comprehensive reform of the country's planning system in sixty years. It was intended to bring in a much more inclusive and efficient planning system to improve community involvement, support the economy, and help it grow in a sustainable way. In Northern Ireland, specific local legislation to control the development of land was first introduced in the early 1930s and for forty years local government administered the planning system. In 1973 local government was reorganised and when the provisions in the Planning (NI) Order 1972 were commenced, the Ministry of Development became the planning authority for Northern Ireland in place of the local planning authorities. Responsibility for planning control was subsequently transferred to the Department of the Environment, which is now responsible under the Planning (Northern Ireland) Order 1991 for planning matters. In early December 2010 a draft Planning Bill was agreed by the Executive and was introduced to the Northern Ireland Assembly. The Bill will provide for the transfer of the majority of planning functions from central government to district councils within a timetable to be agreed by the Executive. It also brings forward a number of reforms to the Northern Ireland planning system.

Types of planning consent

Throughout the UK, there are two types of planning permission – 'outline' and 'full permission' – which can be applied for from the planning authority. It should be noted that anyone can apply for planning permission even if it is not the owner of the land, but the person must inform all other parties as well as those with seven years or more left on a lease.

Outline permission is generally used to find out, at an early stage, whether or not a proposal is likely to be approved by the planning authority before any substantial costs are incurred. If granted outline permission, approval of the details is still required before construction can commence. Outline permission is valid for up to three years; if a detailed application for full approval has not been made within that time frame then the outline permission will expire. Full permission is when the client makes a full application for either a new building or change of use. If granted, work should start within three years otherwise the client will have to reapply.

When an application is made to the local authority planning department, the submission is assessed on a number of factors. These include:

- Whether the scheme fits within the authority's development strategy for the area.
- The size, number, layout and siting of the buildings.
- The external appearance and height of the buildings in relation to other buildings and the surrounding area.
- The proposed use of the buildings and the impact that they would have on the landscape and transportation issues – means of access, effects on parking and traffic.

A full planning application must be accompanied by details of the proposed work, a plan of the site and a fee. This application needs to be submitted in triplicate. There is also a requirement to complete a certificate to show that you own the land/building or that you have informed the person who does. The fee, which is paid to the local authority planning department, depends on the floor space or change of use.

Applications to the planning authority are generally decided at a meeting of the planning committee or under 'delegated powers' whereby a planning officer can determine the outcome of a planning application without the requirement for the application to be put before the planning committee. These planning committee meetings are open meetings and the applicant together with professional advisers can be invited to attend. In some cases applications are delegated to the planning officer without the need for a committee meeting.

Prior to a decision being made, the planning authority undertakes a consultation with the adjoining neighbours and publishes details of the application on its website. This enables members of the public to comment on the scheme and in some instances their views will be taken into account when a decision is made on the application. There is a statutory time limit within which most applications should be determined of eight weeks from submission of the formal application, but for larger complex schemes it may take longer.

If the planning authority is unable to decide upon your application within eight weeks, they may write to you requesting an extension of time. If you refuse, then you are able to make an appeal on ground of 'non-determination'. If the planning authority offers a refusal, with reasons as to why attached, or the terms of conditions are unacceptable, you can appeal to the Secretary of State or the equivalent in Scotland, Wales and Northern Ireland.

There are circumstances when planning consent is required and is granted by Statutory Instrument, a process where the UK Parliament delegates powers to government ministers to create legislation. These are commonly known as 'permitted development rights'. In October 2008, these rights were clarified and extended to cover more building operations. As a result, some projects can be carried out without express planning permission, as long as they meet certain criteria like dimensions and position. These include:

- Extensions and conservatories
- Loft conversions
- Roof alterations
- Fitting of solar panels
- Patios and driveways.

It should be noted that these 'permitted development rights' are restricted in conservation areas, national parks, areas of outstanding natural beauty and the Norfolk or Suffolk Broads and in buildings which are recognised as 'listed buildings' (we will discuss listed buildings in the following chapter).

Permitted development

In 2010, central government introduced 'The Town and Country Planning (General Permitted Development) Order (GPDO)'. This superseded the 1995 and 2005 orders, which classified the types of development that could be carried out without planning permission, otherwise known as 'permitted development'. The new order expands the scope of permitted development for non-domestic premises. The principal changes to be introduced by the 2010 amendment order were:

- Planning consent is not required for existing industry and warehouse developments which want to construct new buildings up to 100 m².
- Planning consent is not required for schools, colleges, universities and hospitals that want to construct new buildings or extension up to 25 per cent of the gross floor space of the original building, or 100 m², whichever is the lesser.
- Planning consent is not required for office buildings which want to extend their floor area up to 25 per cent of the gross floor space of the original building, or 50 m², whichever is the lesser.
- Planning consent is not required for shops or financial and professional services establishments that want to extend their premises up to 25 per cent of the gross floor space of the original building or 50 m², whichever is the lesser.

All of these permitted development rights are subject to certain constraints, designed to minimise impacts on neighbours and the wider environment. In October 2008, significant amendments to the GPDO came into force in respect of Part 1 'householder permitted development (PD) rights'. These replace the current PD rights for householders which have remained essentially unchanged since 1995.

The history of building regulations to date

The first recorded 'building regulation' was in 1189 where thin party walls and badly sited privies were some of the conditions that were tackled within the City of London, but it wasn't until the Great Fire of London when significant building regulations were established. Following the Great Fire of London in 1666, the Rebuilding Act of 1667 first introduced the concept of building control into London. By the mid-eighteenth century the model of building control had been established in many cities throughout England and in Scotland the Burghs developed their building regulations which were operated by the Dean of Guild Court. In some cities, like Bristol in 1778 and 1840 and Liverpool in 1825 and 1842, similar Acts were passed.

Over the years that followed, the legislation covered by Building Control evolved and grew. In 1936 it was standardised as the Building Regulations under the Public Health Act 1936, which itself grew from the 1848 Public Health Act. Within this Act a series of controls regarding the construction and condition of buildings was developed. This was also the time when British Standards were created, but these were not made mandatory as the Building Regulations were. As a result, many local authorities continued to use their own standards or bye-laws. Some of these bye-laws are still in force today.

In 1959, Scotland was the first country to adopt national regulations via the Building (Scotland) Act. By the 1960s the government had repealed the powers enabling local authorities to make building bye-laws and in 1965 they created the first national compulsory Building Regulations 1965 in England and Wales, which were now distinct from the Public Health Act. In 1984 the UK Parliament passed a 'Public General Act' which formed the Building Act 1984. Over the years that followed, the Building Regulations were adapted to meet the changing needs of society to their current status as secondary legislation and accumulation of technical guidance, which themselves set the minimum standards that are considered reasonable in the design and construction of a wide range of buildings.

What is covered by the Building Regulations?

A 'building' is defined in the 1984 Building Act in very wide terms. A building is 'any permanent or temporary building and, unless the context otherwise requires, it includes any other structure or erection of whatever kind or nature (whether permanent or temporary)'. Equally the term 'structure or erection' includes a vehicle, vessel, hovercraft, aircraft or other movable object of any kind in such circumstances as may be prescribed by the Secretary of State. The result of this definition is that many things, which would not otherwise be thought of as a building, may fall under the Act – fences, radio towers, silos, air-supported structures and the like. Thankfully, there is a more restrictive definition of 'building' for the purpose of the Building Regulations.

The Building Regulations apply only to 'building work' or to a 'material change of use', when an office building is converted into a block of residential flats for example. Work which relates solely to maintenance and repair is not normally controlled. Nevertheless, window replacements are controlled and can either by carried out by a registered competent person (see below) or under the scope of 'material alterations' as referred to within the means of 'building work'.

Under the regulations the meaning of building work refers to:

- The erection or extension of a building. This could include a single- or two-storey rear or side extension or a loft conversion with an alteration to the roof by the installation of a dormer window. There are some extensions which are considered exempt from the Building Regulations and these will be discussed further in the section below that covers 'Exemptions'.
- The provision or extension of a controlled service, or fitting in or connection to a building. A controlled service or fitting would include the following: fixed heat producing appliances, baths, toilets, sinks, washbasins, controlled heating systems and domestic electrical installations. If it is the intention to replace a unit like for like, then an application would not be required. This tends to happen when a client wants to change an old bathroom suite for a new suite, without changing the actual position of the units. It would cover the installation of a cloakroom under the stairs or the creation of an en suite bathroom to an existing bedroom.
- The material alteration of a building, or a controlled service or fitting. Consent is required where an alteration such as an extension, loft conversion or conversion into flats is undertaken. Some people believe that planning and building control is only required if you change the external appearance of a property, but in reality consent from a Building Control Body is required when you undertake internal alterations within a property. This could include removal of a chimney breast, creation of a through-lounge, altering the existing layout of the dwelling to create an open-plan arrangement, converting a basement for habitable use, or the underpinning of the existing foundations. It is true to state that works which relate to maintenance and repair, such as replacement of, say, a bathroom suite in a like-for-like position or the laying a new wooden floor covering, will not require consent.
- Creation of a basement or indoor swimming pool to the property.
- Repositioning of the staircase to the upper floors.
- Creation of new windows and doors.
- Work required for a material change of use where the building:

  - is used as a dwelling, where previously it was not
  - contains a flat, where previously it did not
  - is used as a hotel or boarding house, where previously it was not
  - is used as an institution, where previously it was not
  - is used as a public building, to which members of the public have access, where previously it was not
  - is not an 'exempt building', where previously it was
  - contains a greater or lesser number of dwellings that it did previously, such as when a house is converted into flats.

- The insertion of cavity wall insulation material. This tends to be undertaken by companies that fill the clear cavity of an existing house with thermal insulation material. It should be noted that this process should not be carried out to a party wall. As an example an owner decided to fill the party wall. The neighbour returned home from shopping to find the living room covered with blown mineral fibre insulation.
- Work involving the underpinning of a building. Underpinning is the process of supporting the existing foundations of a property by extending the depth of foundations through

underpinning This is normally undertaken as a result of subsidence where the building has moved due to poor soil conditions or a change in the soil conditions as a result of tree root activities for example. You may also need to underpin if you decide to add another storey to the building, either above or below ground level, and the depth of the existing foundations is inadequate to support the modified building or load (weight). Underpinning can be undertaken in a number of ways, either by physically digging the ground under the foundations to an adequate depth and then filling the void created with concrete or by augering a pile casing through the existing foundation and filling the pile with concrete.

Exemptions from the Building Regulations in England and Wales

Under the recent changes to the Building Regulations in October 2010, there are a number of domestic buildings which are exempt from the requirements contained within the guidance documents. These include:

- Detached single-storey buildings like garages and sheds with an internal floor area of not more than 30 m² and which contain no sleeping accommodation. In addition, this building should be more than one metre away from the boundary of its curtilage or, if not, should be constructed substantially of non-combustible material. This means that the walls should be brick or block and the roof can be felted with three layers with an 'AA', 'AB' or 'AC' surface spread of flame rating or traditional clay tiles or slates on timber joists. Timber window and door frames and wooden fascias are disregarded. Nevertheless, any drainage for a toilet needs to comply with the requirement of Part H, 'drainage'; any hot and cold water system must comply with Part G, 'sanitation, hot water safety and water efficiency'; and any fixed electrical installation must comply with the requirement of Part P, 'electrical safety'. In these circumstances consent would be required.
- **Extensions (which are not detached),** for example a conservatory, porch, covered yard or covered way or even a carport (open on at least two sides), where the floor area of that extension would not exceed 30 m². A conservatory is defined as a structure where not less than three-quarters of the area of its roof and not less than one-half of the area of the external walls are made of translucent material like glass or plastic. The only requirements are that in the case of a conservatory or porch, which is wholly or partly glazed, the glazing satisfies the requirements of safety glazing and any fixed electrical installation complies with the requirement of Part P ('electrical safety') of the Building Regulations. As mentioned above, if the porch was to incorporate a toilet, the drainage would require consent. Equally, if the existing front door behind the porch was removed such that the porch door would become the front door, then this would not constitute an exempt proposal and the porch would require consent. The same principle applies where you enter a conservatory from a habitable room, without having to pass through a door between the two. If this is the case, then consent would be required.

There are also a number of non-domestic buildings which are exempt, these include:

- A building which people don't normally enter unless it is to inspect or maintain fixed machinery or plant, except where the building is one and half times in height from the boundary of another building where people are likely to go.
- Greenhouses or agricultural buildings, except where the building is one and half time in height from a building which contains sleeping accommodation or the building is used for retailing, packing or exhibiting.

- A temporary building which is not intended to remain where it has been erected for more than 28 days.
- Ancillary buildings, which are used in connection with the disposal of buildings or building plots on that site or which are on a construction site and used by staff on a daily basis but contain no sleeping accommodation.

Exemptions from the Building Warrants in Scotland

Within the Building (Scotland) Regulations and for the purpose of Section 8(8) of the Building (Scotland) Act 2003, Regulation 5 and Schedule 3 specifies what work can be done without the need to obtain a Building Warrant, subject to those works still complying with the requirements of the regulations. Further guidance can be found within Section 0.5 of the Technical Handbook published on the Building Standards Section of the Scottish Government's website, http://www.sbsa.gov.uk/tech_handbooks/tbooks2009.htm.

Exemptions from the Building Regulations in Northern Ireland

Guidance is given within Regulation A5 and Schedule 1 of the Building Regulations (Northern Ireland) 2000. Within these regulations similar exceptions to those mentioned in the England and Wales regulations are included, but additional structures such as air-supported structures which do not exceed 15 m in length or diameter, have an alternative escape route and are solely used as dwellings, tents or mobile accommodation are also exempt. For further guidance I would recommend you view the regulations at http://www.legislation.gov.uk/nisr/2000/389/regulation/A5/made.

Guidance documents

Unlike planning legislation, the building regulation process addresses not whether the applicant should be allowed to build, but whether the proposed construction complies with the Building Regulations, which also contain a list of requirements, referred to as Schedule 1 of the Building Regulations or commonly known as 'Approved Documents' within England and Wales. The requirements within the Approved Documents are expressed in broad, functional terms in order to give designers and builders the maximum flexibility in preparing their plans. They indicate the minimum standards which are required to ensure compliance and contain practical and technical guidance on ways in which the requirements can be met. These Approved Documents cover subjects such as structure, fire safety, drainage and energy conservation to name but a few. In addition to Schedule 1, the Building Regulations also contain Schedule 2, which lists buildings and extensions that are exempt from Building Control consent.

Even though the Approved Documents outline legal requirements and give practical and technical guidance on how to meet the regulations, you are not obliged to use any particular solution in the Approved Documents if you prefer to meet the requirements of the regulations in another way. However, following the guidance in the Approved Documents will provide evidence to show that you have tried to comply with the regulations. If you do not follow the guidance then it will be for you to demonstrate by other means that you have satisfied the requirement of the regulations. Nevertheless, it is important that all building work is carried out so that after it has been completed it will comply with the relevant requirements of Schedule 1 or, where it does not comply with any such requirement, is no more unsatisfactory in relation to that requirement than before the work was carried out.

Within England and Wales, the Approved Documents cover the following aspects:

- Approved Document A Structure – This document covers the requirements with regard to structural safety. It also makes reference to other documents which offer guidance on how to structurally design a building.
- Approved Document B Fire safety – Volume 1: Dwelling houses.
- Approved Document B Fire safety – Volume 2: Buildings other than dwelling-houses. Both Volumes 1 and 2 cover the requirements with regard to fire safety and are broken into various sections:

  - Section B1 covers means of escape in case of fire. This includes fire detection alarm systems, emergency lightings, travel distances. For non-domestic buildings it includes design for both horizontal and vertical escape.
  - Section B2 covers internal fire spread in connection with linings. This includes fire resistance of the walls, ceilings and thermoplastic material.
  - Section B3 covers external fire spread in connection with the structure. This includes the fire resistances of a load-bearing element. The following are classified as a load-bearing element: walls, floors and structural elements which are supporting items – but not a roof, unless it's part of a terrace. It also includes compartmentation and concealed spaces within cavities and issues around fire stopping and the protection of openings. For non-domestic buildings it also includes special provisions for smoke venting of car parks and shopping complexes.
  - Section B4 covers external fire spread. This includes the construction of external walls, the limitation on the amount of glazing allowed on a boundary condition and fire resistance of roof coverings.
  - Section B5 covers access and facilities for the fire and rescue services. This includes the design requirements for vehicle access, and for non-domestic buildings it includes fire mains, access to the building for firefighting personnel and venting of heat and smoke from buildings.

- Approved Document C Site preparation and resistance to contaminants and moisture – This covers the requirements with regards to site preparation and resistance to contaminants and moisture. This includes clearance or treatment of unsuitable material and resistance to contaminants and sub-soil drainage. This also includes guidance on the resistance to moisture of floors, walls and roofs, and radon protection.
- Approved Document D Toxic substances – This covers the requirements of dealing with toxic substances and tends to primarily feature cavity wall insulation which is spray applied.
- Approved Document E Resistance to the passage of sound – This covers the requirement of sound insulation within a building, but does not include requirements for external noise affecting a building, such as traffic etc. The document states the sound insulation value which should be achieved between dwelling unit floors and walls as well as within the common areas.
- Approved Document F Ventilation – This covers the requirement of adequate ventilation of a building. This includes natural, mechanical and passive ventilations.
- Approved Document G Hygiene – This covers the requirement of adequate sanitation, hot water safety and water efficiency and sets out the 'water calculation methodology' for assessing the whole house water efficiency of new dwellings, if not design to code for sustainability level 2.

- Approved Document H Drainage and waste disposal – This covers the requirement for foul water, waste water treatment and cesspool, rainwater drainage, building over sewers, separate drainage and solid waste storage.
- Approved Document J Combustion appliance and fuel storage systems – This covers the requirement of heat producing appliances. This includes air supply, carbon monoxide guidance and liquid storage systems.
- Approved Document K Protection from falling, collision and impact – This covers the requirement for stair design, vehicle barrier, and protection from falling. In addition it covers protection from collision with opening windows and protection against impact from and trapping by doors.
- Approved Document L1A Conservation of fuel and power, new dwellings.
- Approved Document L1B Conservation of fuel and power, existing dwellings.
- Approved Document L2A Conservation of fuel and power, new buildings other than dwellings.
- Approved Document L2B Conservation of fuel and power, existing buildings other than dwellings – All the part L documents cover the thermal efficiency required to meet the regulations. For new dwelling houses designers are required to use the Standard Assessment Procedure model (SAP). SAP is a computer program which is used to estimate the carbon emissions from heating, hot water, lighting, the solar gain from windows, the air tightness of a building, together with mechanical services. The result of the calculation produces a dwelling emission rate (DER) which should be better or equal to a target emission rate (TER). Under recent changes to Approved Document L in 2010, there is a requirement for the person who is undertaking the proposed work to provide Building Control with the DER and TER ratings, together with a list of specifications before work on site commences. Within five days of completion they must notify Building Control of the DER and TER actually achieved and state if there are any changes to the design from the previously submitted specifications. For existing dwellings, there is guidance on the thermal insulation requirements needed for walls, floors, roof and even windows.
- Approved Document M Access to and use of buildings – This covers the requirement that both disabled people and others are enabled access to and around buildings and facilities within the buildings.
- Approved Document P Electrical safety, dwellings – This includes the design and installation of electrical systems.
- Approved Document 7 Materials and workmanship – This offers guidance to ensure that any building work shall be carried out with proper materials and in a workmanlike manner.

Similar requirements are provided for in Scotland and Northern Ireland and work in the same manner as the Building Regulations for England and Wales. In Scotland, these are referred to as Technical Handbooks, and provide guidance on achieving the standards set in the Building (Scotland) Regulations 2004 and are available in two volumes, for Domestic Buildings and for Non-Domestic Buildings.

- Domestic 0 General
- Domestic 1 Structure
- Domestic 2 Fire
- Domestic 3 Environment

- Domestic 4 Safety
- Domestic 5 Noise
- Domestic 6 Energy
- Non-Domestic 0 General
- Non-Domestic 1 Structure
- Non-Domestic 2 Fire
- Non-Domestic 3 Environment
- Non-Domestic 4 Safety
- Non-Domestic 5 Noise
- Non-Domestic 6 Energy.

Within Northern Ireland the equivalent are known as 'Technical Booklets' and are published by the Department of Finance and Personnel of the Northern Ireland Government in support of some of the technical parts of the Building Regulations. Like the English requirements, they provide construction methods that, if followed, will be deemed to satisfy the requirements of the Northern Ireland Building Regulations.

- Technical Booklet C Site Preparation and resistance to moisture
- Technical Booklet D Structure
- Technical Booklet E Fire Safety
- Technical Booklet F1 Conservation of fuel and power in dwellings
- Technical Booklet F2 Conservation of fuel and power in buildings other than dwellings
- Technical Booklet G Sound
- Technical Booklet G1 Sound (conversions)
- Technical Booklet H Stairs, ramps, guarding and protection from impact
- Technical Booklet K Ventilation
- Technical Booklet L Combustion appliance and fuel storage system
- Technical Booklet N Drainage
- Technical Booklet P Unvented hot water storage system
- Technical Booklet R Access to and use of buildings
- Technical Booklet V Glazing.

Building Regulation consent

To ensure that the proposed scheme complies with the current Building Regulations an application is submitted for Building Regulation consent under which the scheme will be assessed and the construction work on site inspected by a Building Control surveyor. Work which is undertaken without consent can obtain 'regularisation consent' from the local authority Building Control surveyor. This may be obtained by submitting an application and exposing aspects of the work such as the foundations, damp proofing, drainage, structural steelwork and insulation for the Building Control Surveyor to inspect, assess and approve of if found to be satisfactory. Once completed satisfactorily a final certificate or a Regularisation Certificate will be issued. Currently the role of Building Control can be undertaken by either the local authority or an Approved Inspector. It should be noted that Approved Inspectors are not authorised to undertake the functions of Building Control in either Scotland or Northern Ireland.

When using the services of the local authority, there are two routes for obtaining consent. The first is known as a 'Full Plans application', for which plans need to be produced showing all constructional details, preferably well in advance of the intended commencement date of work on site. The second is known as a 'Building Notice Procedure'. In both these cases the application or notice should be submitted to the local authority and should be accompanied by any relevant structural calculations in order to demonstrate compliance with safety requirements on the structure of the building. It is advisable that this be undertaken by a person competent to do so. Regardless of which route is chosen, if work on the scheme is not commenced within 3 years from the date of deposit to the local authority, a new notice for the work will be required.

Under a 'Full Plans application' the scheme will be thoroughly checked by the local authority which is required to pass or reject the plans within a certain time period, or it may add conditions to an approval, with your written agreement. If they are satisfied that the work shown on the plans complies with the regulations, you will be issued with an approval notice within a time period of five weeks or up to two months if you agree to this. This will give you the protection of being able to show that your plans were approved as complying with the Building Regulations. If the building is constructed following these 'approved plans' then the constructed building will comply with the Building Regulations as well. If your plans are rejected, and you do not consider it is necessary to alter them, you will have two options available to you:

- You may seek a 'determination' from the Secretary of State if you believe your work complies with the regulations, but you should apply before work starts; or
- If you acknowledge that your proposals do not necessarily comply with a particular requirement in the regulations and feel that it is too onerous in your particular circumstances, you may apply for a 'relaxation' or 'dispensation' of that requirement from the local authority. You can make such an application at any stage but it is obviously sensible to do so as soon as possible and preferably before work starts. If the local authority refuses your application you may then appeal to the Secretary of State within a month of the date of receipt of the rejection notice.

Under the 'Building Notice' procedure no approval notice is given and theoretically the client can commence work on site 48 hrs after the Building Notice has been submitted to the local authority. Since there is no approval or rejection offered by the local authority, there is also no procedure to seek a determination from the Secretary of State if there is a disagreement between you and the local authority. However, the advantage of the building notice procedure is that it allows you to carry out works without the need to prepare full plans, for minor works for example. Nevertheless you must feel confident that the work will comply with the regulations or you risk having to correct any work you carry out at the request of the local authority following their inspection to ensure compliance. There are limitations to the use of this route since it only applies to extensions of dwelling houses and not for the construction of new dwelling houses, nor for work pertaining to commercial, industrial, office, educational, healthcare or assembly buildings because under the Building Regulations the local authority needs to consult with the fire authority and consequntly requires appropriate plans and at least 15 days in which to undertake the consultations.

Under the Building Act plans are defined as 'includes drawing of any other description, and also specification or other information in any form'. It is possible for the approved plans to be amended following discussion and consent from the local authority.

The role of an Approved Inspector is one which is an alternative to local authority Building Control. The Construction Industry Council (CIC) has been designated by government as a body for approving inspectors. Individual and corporate Approved Inspectors registered with CIC are qualified to undertake Building Control work. When you use an Approved Inspector, their 'Initial Notice' procedure takes the place of the Full Plans application and Building Notice procedures. In addition, Approved Inspectors can grant formal plans certificates to give similar protection to that provided by a local authority Building Control service. The regulations stipulate that any Initial Notice should be served on and accepted by the local authority Building Control Body at least 5 days before the work commences on site. If this time period has not been met, the local authority Building Control service is within its rights to reject the Initial Notice and require the proposed scheme to be controlled by itself.

If you use an Approved Inspector, the information needed to ensure that the work meets Building Regulation standards is a matter for discussion between you and the Approved Inspector who will advise you about what is required. The local authority Building Control will not be involved in the scheme unless there is a contravention of the Building Regulations on site and enforcement action is required. In these circumstances the scheme will revert to the local authority Building Control service for resolution. If this should occur then a 'reversion charge' is payable to the local authority which is equivalent to the relevant Building Notice charge. If necessary, formal enforcement to correct defective uncertified work will be undertaken by the local authority.

Within Scotland the system is slightly different to that in England, Wales and Northern Ireland. The Scottish legislation allows for 'Building Warrants' instead of approvals of either Full Plans or Building Notice. Like England and Wales, a Building Warrant is valid for three years from the date of approval. Unlike England and Wales, all works stated within the Building Warrant should be completed and a request for a final certificate should be made prior to the expiry date. If it is felt that the project will extend beyond the planned end date, an extension to the Building Warrant should be made to the local authority before the Building Warrant expires. This must be undertaken on an approved form detailing the reasons for the extension. This request will be considered and a written response will be issued. It should be noted that the request for an extension is assessed on its own merits and approval is not always granted. Where this occurs, and where works do not complete within the time period, the result may be that an enforcement notice is served.

The assessment procedure with the local authority is similar except that Scotland has the ability for the application process to be quicker if an 'approved certifier' is used. The Building (Scotland) Act 2003 introduced the option to certify the design or construction of building work as 'complying with the Building (Scotland) Regulations 2004'. The certification system is based upon the principle that qualified and experienced building professionals/tradesmen can (without the need for further scrutiny by local authorities) take responsibility for ensuring compliance with the regulations, provided they are employed by reputable companies which operate a system of careful checking. There are two kinds of certifier:

- an approved certifier of design, who can check specific elements of the design before you submit them to the Building Standards department, and
- an approved certifier of construction, who can check specific elements of the building work being carried out.

Currently there are two schemes approved by the Scottish Parliament:

* Certification of Design (Building Structures) operated by Structural Engineers Registration Ltd (SER LTD).
* Certification of Construction (Electrical installations to BS 7671) operated by SELECT (Scotland's trade association for the electrical, electronic and communications system industry) and NICEIC (National Inspection Council for Electrical Installation Contracting).

An approved certifier of design can check your plans before you apply for a Building Warrant and issue you with certification under the scheme to show that the plans comply. This should speed up the application process, as all the Building Standards department needs to do on receipt of your application is check the certificate. You'll then get a discount on the warrant fee, although you will have to pay the certifier for issuing the certificate. When your building project is being carried out, you can ask an approved certifier of construction to issue a certificate to say the work complies with building standards. This should speed up the completion certificate process. In addition, if you let the Building Standards department know you will be doing this before you start work, you will get a refund when you submit the completion certificate and the certificate of construction.

Unlike England and Wales, it is an offence in Scotland to commence building operations that require the benefit of a building warrant prior to a warrant being granted. It is also an offence to occupy a building without the benefit of a certificate of completion.

Competent person

Competent person schemes were introduced by central government to allow individuals and enterprises to self-certify that their work complies with the Building Regulations as an alternative to submitting a building notice or using an Approved Inspector. The powers used to set up schemes are contained in Schedule 1, paragraph 4(a) of the Building Act 1984. The current schemes have been set up under the Building Regulations 2000.

As a result of these schemes no Building Control fees are payable and on completion of the project the client is issued with a certificate to show compliance with the Building Regulations. For work that falls under these schemes, the competent person informs the respective scheme organiser, who in turn informs the local authority that the work has taken place such that they can make a record on the system against the property address. An example of such a scheme is the installation of new windows under the FENSA scheme, which is overseen by the Glass and Glazing Federation.

Assessment

Regardless of who undertakes the functions of Building Control for the scheme, a major part of the process is the site inspection. These inspections are undertaken to ensure that the work carried out follows the approved plans or is constructed in accordance with the Approved Documents. The Building Control Body overseeing the scheme should be notified before construction work such as foundations, drainage, steelwork erection, construction of walls or floors commence. In addition, notice of one clear working day must be given for critical inspections like foundations and drainage. Likewise, notice should be given upon the completion of the scheme and prior to occupation. Prior to completion a number of additional certificates should be offered to the Building Control Body. These include:

- Air tight test certificate for new premises.
- Sound test results for new premises including altered premises.
- Test certificates for electrical installations within dwellings.
- Thermal efficiency calculations for premises if amendments to the original plans have taken place which may affect either the Standard Assessment Procedure (SAP) or Simplified Building Energy Model (SBEM) calculations.
- Commission certificates for certain services within the scheme.
- Fire Safety Management plan for non-domestic premises.
- Energy performance certificates.
- Water calculation for newly constructed dwelling home which does not meet Level 2 Code for Sustainability.

Once all these certificates are provided and there are no outstanding matters on site, the Building Control Body will issue, without the need for further payment, a completion certificate which will be evidence (but not conclusive evidence) that the relevant requirements have been met.

So do we need the Building Regulations?

Some may say that there are too many legislative requirements within the construction industry, which might appear to hamper people who wish to build or alter their properties; others may see the regulations as a revenue earner for those who issue consents. Whatever negative views people may hold, the legislation is there to ensure that buildings are constructed or altered in a way that compliments the surrounding environment as well as being built in an appropriate and safe manner.

To provide guidance through the various requirements, construction professionals can provide advice and solutions on: boundary conditions, party wall notices, planning consent, Listed Building consent and Building Control consent. All of these topics are covered in the chapters that follow.

So what does this professional person look like – well ideally they should be a member of a professional organisation like the Association of Building Engineers (ABE) <www.abe.org.uk>. This is a body that represents a mixed group of professionals who practice in the fields of building engineering, building surveying, Building Control surveying, building management, architecture and planning. Other professional organisations include the Royal Institute of Chartered Surveyors (RICS) <www.rics.org>, the Royal Institute of Chartered Architects (RIBA) <www.riba.org>, the Chartered Institute of Architectural Technicians (CIAT) <www.ciat.org.uk> and, finally, the Institute of Structural Engineers (ISE) <www.istructe.org>.

In some cases where the scheme is complex or large, the client may consider it appropriate to instruct a professional designer or project manager to undertake the complete construction package from the design stage to seeking tenders for the construction and advising the client appropriately. Such a person would need to ensure that all the necessary statutory consents were applied for and granted prior to the commencement of the scheme.

Environmental Health

Environmental Health officers (EHO) are mostly employed by the local authority and deal with a range of issues affecting health. Part of their role is to monitor and control pollution levels and educate the public in protecting the environment. The EHOs keep a close check on the levels of air, water, soil and noise pollution in their area, and communicate this information in a way that is meaningful to the public. EHOs are responsible for monitoring the levels of airborne emissions from small and medium sized industrial processes and can take action to cut levels if necessary. Noise can also be a pollutant, arising from such diverse sources as industrial fans, late night parties or burglar alarms, and EHOs have to be impartial but find a practical solution that suits everyone. The need to use and reclaim land as effectively as possible means that attention is being turned to identifying land which is contaminated by refuse, toxic or dangerous waste and a potential danger to health. This aspect of an EHO's work provides an example of how they work alongside other professionals, such as land developers, architects and engineers, to provide a multi-disciplinary approach to a problem.

The process of construction, whether internal or external, creates a number of issues which may have an effect on neighbouring properties and members of the public. Consequently, there is some legislation that is enforced by local authority Environmental Health departments. This includes:

- noise nuisance
- dust nuisance
- smoke nuisance
- light nuisance
- waste disposal
- vermin
- contaminated land.

Under the Control of Pollution Act 1974 local authorities can serve a notice imposing requirements as to how construction work should be carried out with regard to minimising noise and the Environmental Health department sets noise limits, taking into account the character of the local area.

Section 61 of the Control of Pollution Act 1974 states that large developments should apply for approval of work before they commence. The developer must submit a method statement detailing the proposed hours of work and type of plant to be used as well as any noise control methods. Failure to apply under Section 61 will enable the local authority to serve notice in order to control noise and they may prosecute for failure to comply with the notice.

Under the Environmental Protection Act 1990, there are requirements to deal with the generation of dust as a result of either construction or demolition. The Act recommends that good practice be followed to minimise the emission of dust that would impact on the air quality of those near the site in question. Example of good practice techniques could be:

- using plant with dust suppression
- material drop heights kept to a minimum
- positioning of stockpiles to minimise effect of wind
- dust sheets over surface of stockpiled material
- spraying water over dusty areas to damp down, particularly in windy and/or hot weather
- wheel washing facilities at site exits
- road sweeping.

Under both the Environmental Protection Act 1990 and Pollution Prevention and Control Act 2000 there are requirements to ensure that there is limited nuisance from sites as a result of rubbish burning and the generation of smoke. The requirements clearly state that:

* Demolition and construction waste should not be burned.
* If you wish to burn waste you must register an exemption with the Environment Agency and only waste wood or plant material generated as a result of demolition work may be burned, however you must burn the material on the land where it was produced, and this can be up to a maximum of 10 tonnes per 24-hour period.
* There must be no burning of any waste on site that causes smoke nuisance or results in dark smoke.

Under the Environmental Protection Act 1990, Clean Neighbourhoods and Environment Act 2005 and Site Waste Management Plans Regulations 2008 there is legislation to deal with the waste that is generated by either construction, demolition or renovation works. Disposal of construction waste, which is considered commercial waste, should be via a Registered Waste Carrier and if the construction project is over £300,000 (excl. VAT) the contractor should have a site waste management plan (SWMP). There is also a requirement to limit the nuisance caused to neighbouring premises by lighting on a construction site. As a result there are recommendations that construction lights are orientated away from residential windows or that they are turned off outside normal site working hours.

Under the Prevention of Damage by Pests Act 1949 landowners have a general duty to keep their land free of rats and mice. It is recommended that sites are clear of refuse, food waste and other material capable of providing harbourage to rodents. During any demolishing of buildings, there is a requirement to ensure that rodents do not escape from the drainage network. This aspect is also covered within Section 81 of the Building Act 1984 which enables the local authority to issue a counter notice on the person who has informed the authority that they wish to undertake the demolition of a building.

Under the Environmental Protection Act 1990, and The Environment Act 1995, it is the developer's responsibility to ensure that the proposed site is free from contamination and is suitable for the proposed use. If it is found that the land is defined as contaminated and no appropriate action has been taken, then the developer is liable and action will be taken by the local authority.

Poor housing conditions: Housing Health and Safety Rating System (HHSRS)

The HHSRS was introduced into the UK by the Housing Act 2004 and is the current statutory assessment for housing. The HHSRS covers all residential properties including owner-occupied, privately rented and social housing properties. It is a system that is used by the local authority, which will inspect the property to assess existing housing conditions using a risk assessment process. HHSRS uses a formula to produce a numerical score based upon an inspection of the whole property. The assessment will show the presence of any serious (category 1) hazards and other less serious (category 2) hazards and will consider:

(a) the likelihood, over the next twelve months, of an occurrence that could result in harm to a member of the vulnerable group;
(b) the range of potential outcomes from such an occurrence.

The HHSRS is concerned with 29 hazards which are divided into four groups:

Physiological requirements

- Damp and mould growth
- Excess cold
- Excess heat
- Asbestos and manufactured mineral fibre
- Biocides
- Carbon monoxide and fuel combustion products
- Lead
- Radiation
- Uncombusted fuel gas
- Volatile organic compounds.

Psychological requirements

- Crowding and space
- Entry by intruders
- Lighting
- Noise.

Protection against infection

- Domestic hygiene, pests and refuse
- Food safety
- Personal hygiene, sanitation and drainage
- Water supply for domestic purpose.

Protection against accidents

- Falls associated with baths
- Falling on level surfaces
- Falling associated with stairs and steps
- Falling between levels
- Electrical hazards
- Fire
- Flames and hot surfaces
- Collision and entrapment
- Explosions
- Position and operability of amenities
- Structural collapse and failing elements.

If a local authority discovers serious category 1 hazards in a home, it has a duty to take the most appropriate action. This may include powers to prohibit the use of the whole or part of a dwelling or restrict the number of permitted occupants. Where hazards are modest they may serve a hazard awareness notice to draw attention to a problem. Where an occupier is at immediate risk, the authority can take emergency remedial action. Failure to comply with a statutory notice could lead to a fine of up to £5,000. If a person feels that the local authority's assessment is not correct, they can challenge any enforcement through the Residential Property Tribunal.

Highways authorities

The Highways Act 1980 was created primarily to give power to the highways authorities to construct roads, issue compulsory orders, create footpaths and bridleways, as well as pay compensation to people whose properties are affected by new roads, usually as a result of noise. Highways may be constructed by the highway authorities under Section 24 of the 1980 Act. The usual practice is for the land to be acquired by the highway authority under the compulsory purchase powers in Section 239 of the 1980 Act. Local planning authorities may also grant planning permission for the construction of highways (often as part of wider development proposals). Prior consultation between the planning officer, tree officer and highway engineer will ensure that the trees are protected and that any new trees which are planted in or near the highway do not hinder reasonable use of the highway.

Precautions to be taken by builders

The Highways Act 1980 and the Roads Scotland Act 1984 contain a number of provisions concerning precautions to be taken by people doing certain works in or near streets or highways. The following sections refer to the Highways Act 1980.

○   Section 168 – Building operation affecting public safety: This states that if, in the course of carrying out any building operation in or near a street, an accident occurs which endangers a person in the street, the owner of the land or building is liable to a fine in addition to any other liability they incur. It is a defence that they took all reasonable precautions to avoid danger or that the offence was due to the act or default of another person in which case that other person may be charged with an offence under the section.

○   Section 169 – Control of scaffolding on highways: This states that no person may, in connection with any building, demolition work, repair, maintenance or cleaning of any building, erect or retain over a highway any scaffolding or other structure that obstructs a highway, without a licence in writing issued by the highway authority. There is an appeal procedure against the refusal to grant a licence to the magistrates' court. A person to whom a licence is issued must ensure that the structure is adequately lit at all times between half an hour after sunset and half an hour before sunrise and comply with the terms of the licence and any direction of the highway authority.

○   Section 170 – Control of mixing mortar etc. on highways: This states that it is an offence to mix or deposit on a highway any mortar or cement or other substance which is likely to damage the surface of the highway or damage any drains or sewers connected with the highway. This section does not apply to any mixing or deposit in a receptacle or on a plate which prevents the substance from coming into contact with the highway and neither from entering any drain or sewers, nor to work carried out by highway authorities or statutory undertakers.

○   Section 171 – Control of deposit of building materials, rubbish, etc. and making temporary excavations in streets maintainable at the public expense: This states that the consent of the highway authority is required for any of these activities and it may be granted subject to such condition as the authority sees fit. A person aggrieved by a refusal of consent or any conditions may appeal to the magistrates' court. Any obstruction or excavation must be properly fenced and, during the hours of darkness, properly lit.

○ Section 172–173 – Hoarding to be set up and securely erected during buildings: This states that a person proposing to erect, take down, alter or repair a building in a street must before beginning the work erect a close-boarded hoarding or fence to the satisfaction of the 'appropriate authority', i.e. the local authority. The obligation may be dispensed with if the appropriate authority consents. The authority may impose requirements as to a covered platform and handrail outside the hoarding, lighting during hours of darkness and the removal of the hoarding. There is an appeal to the magistrates' court against a refusal of consent or any requirement. The hoarding must be securely erected.

○ Section 174 – Erection of barriers and traffic signs, guarding and lighting of works and protecting and shoring up buildings by persons carrying out works in streets: This states that it is an offence for any person executing works in any street, other than the street where the construction work is being carried out.

○ Section 139, 140 and 140A – Control and removal of builders' skips: This states that a builder's skip may not be deposited on a highway without the permission of the highway authority, which may be granted subject to conditions as to siting, dimensions, lighting and guarding. The skip must be clearly and indelibly marked with the owner's name and his address or telephone number. The highway authority or law enforcement authority may require the owner of a skip to remove or reposition it or may themselves remove or reposition it and recover the expense from the owner.

○ Section 74 – Building Lines: This states that the highway authority has the power under the Act to prescribe a frontage line for building on either or both sides of a highway maintainable at the public expense for which they are responsible. The effect of prescribing a building line is to prohibit new building, other than a boundary wall or fence, nearer to the centre line of the highway without the consent of the authority. The authority may give consent for the proposed time period and on such conditions as it deems expedient. Conditions are binding on successive owners and occupiers. Compensation is payable to any person whose property is injuriously affected by the laying down of a building line provided the claim is made within six months.

Permits

For anyone other than a public utility or their appointed contractor work is not permitted on the public highway without a permit. Permits must be issued for the following:

• Erection of scaffolding or hoarding on the public highway
• Placing of skips or storing of materials on the highway
• Operation or storing of plant on the highway
• A temporary vehicle crossing across the footway (without excavation)
• Construction of a temporary vehicle crossing across the footway (with excavation)
• Construction of a permanent vehicle crossing across the footway
• Excavation in the public highway to connect a sewer
• Any other excavation in the public highway.

If a permit is not taken out, a default charge will be made, plus the normal cost of the permit.

Environmental agencies

The Environment Act 1995 established the Environment Agency and Scottish Environ-mental Protection Agency (SEPA) as the regulatory bodies for contaminated land, control of pollution and conservation or enhancement of the environment and fisheries. The pur-pose of the agencies was 'to protect or enhance the environment, taken as a whole' so as to promote 'the objective of achieving sustainable development'. Within Northern Ireland it is the Department of Environment.

The Environment Agency's involvement within the construction industry is limited to the areas of waste, contaminated land and potential risk to the water source. In April 2008 new regulations were introduced, which meant that any construction project in England costing over £300,000 would need a site waste management plan (SWMP). This covers new build, maintenance, and alteration or installation/removal of services such as sewerage, water.

A SWMP sets out how building materials, and resulting waste, are to be managed during the project. The SWMP's purpose is to ensure that:

- building materials are managed efficiently
- waste is disposed of legally, and
- material recycling, reuse and recovery are maximised.

New developments may be constructed on contaminated land; such contaminates may range from naturally occurring substances in the land, for example arsenic, through to con-tamination caused by industrial processes, or even criminal activity. Where work is proposed to such land then there may be a requirement to clean up or undertake remediation work on the site. This could be requested through the planning process and controlled by both Building Control on site and Environmental Health and the Environment Agency off site due to the disposal of material. The most serious sites are dealt with through the statutory contaminated land regime found within the Environmental Protection Act 1990. Within England and Wales, only those sites that are causing or have the potential to cause signifi-cant harm are formally classed as 'contaminated'. The regulators of such sites are either local authorities or, in the most serious cases, the Environment Agency (in England and Wales) or SEPA (in Scotland). Different rules apply in Northern Ireland.

Laws relating to the pollution of water courses have been traced back to the late four-teenth century, when it was illegal to pollute the rivers. In the nineteenth century this law was updated under the River Pollution Prevention Act 1876. This was subsequently amended by the River Boards Act 1948. This legislation established the River Boards, which had the power to deal with the supply of wholesome water and which oversaw the sewage function. These River Boards worked in an isolated manner, overseeing the legislation within their area, and it was not until the River (Prevention of Pollution) Act 1961 that a more coherent procedure for controlling and limiting the pollution of water through the country was provided. Subsequently, the Control of Pollution Act 1974 was used to enforce the existing controls as well as create new controls, and in 1989 the Water Act established the National Rivers Authority as the main regulatory body for water pol-lution. The role and functions of this body were later transferred to the current Environ-ment Agency, which now primarily oversees the regulatory functions of water pollution in England and Wales following the Environment Act 1995. Scotland and Northern Ireland have equivalent regulators.

The Water Act 1989 itself has been replaced by the Water Resources Act (WRA) 1991, which consolidated existing water laws. Under this Act, discharges to controlled waters, namely rivers, estuaries, coastal waters, lakes and groundwater, are regulated. As a direct result, it is a criminal offence under Section 85 of the Act to release discharges into controlled waters without proper authority from the Environment Agency or Scottish Environment Protection Agency (SEPA). If caught, polluters pay for the environmental costs of their discharges into controlled waters. Under Section 85 there are two categories: 'causing' the offence and 'knowingly permitting' the offence to occur, but refusing to act to stop it. Drainage of water or trade effluents into the sewage system is dealt with by other legislation including the Water Industry Act of 1991.

It is an offence under Section II of the Clean Air Act 1993 to burn anything on an industrial or trade premises which is likely to cause dark smoke. In addition, under the Environmental Protection Act 1990, Part III, nuisance legislation applies where a nuisance is defined as something which unreasonably interferes with someone else's enjoyment of their home or garden. If it is confirmed that a nuisance does exist, a legal document called an 'Abatement Notice' can be served on the person responsible, requiring that the nuisance is stopped. If the 'Abatement Notice' is ignored, then any further nuisance will be an offence. The Environment Agency also has the powers to deal with the burning of waste on trade premises. Under Part II of the Environmental Protection Act 1990, you have a duty of care concerning the disposal of trade waste. Section 33 of this Act states that a person shall not dispose of controlled waste in a manner likely to cause pollution of the environment or harm to human health. Any burning of waste is therefore an offence under this legislation.

Sewage undertaking

During the process of major development, consultation should be undertaken with the relevant water authority for the area with regard to sewage connection. During the process of a small constructional development, such as the erection of a single-storey extension, there is the possibility that the proposed foundation designs may come across various drainage runs. Drainage pipes which are commonly used to discharge foul and storm water from buildings are classified as one of the following:

- Private drain – a drain serving single premises.
- Private sewer – a drain serving two or more premises.
- Public sewer – a drain serving two or more premises owned or adopted by the sewerage undertaker.

The majority of public sewers run under public roads, however in some instances they can be found to cross private land. If the sewer to your property was built before October 1937 then it is also considered to be a public sewer under the provisions of Section 24, of the Public Health Act 1936. Within Approved Document H4 of the Building Regulations it offers guidance to the construction, extension or underpinning of a building over or within three metres of the centre line of an existing drain, sewer or disposal main shown on the Statutory Sewer Map. Copies of the sewer records maps are held by the sewerage undertaker under Section 199 of the Water Industry Act 1991 and by local authorities.

When making a Building Regulation submission to the local authority, building over or within 3 metres of a public sewer up to 225 millimetres in diameter will require a Full

Plans submission under the regulations; a Building Notice cannot be accepted in these circumstances. As part of the process, Building Control has to consult with the water authority. The application will only be approved if the works comply with the requirements of Approved Document H4. When submitting details to an Approved Inspector they would require proof of water authority consent before approving the application.

2 Listed buildings and conservation areas

Today, there is a very strict regime of 'statutory control' in place that governs the construction of almost all types of buildings. When a new building is to be built, it is now well known that there will be a need for applications to be made for appropriate planning permission and also Building Regulations consents, and there are a range of penalties if these permissions and consents are not obtained. These can be as draconian as requiring the removal of a property that has not been given the right number of green lights, or even simple fines. Overall, those involved with drafting the necessary 'planning' legislation see themselves as defenders of the public interest, and the rules are implemented to try to encourage a positive effect in a local environment. Those in the Building Regulations sector have a slightly different bent about them, they are more concerned with 'quality control' – the rules 'is' the rules and woe betide you if what they see as the cornerstone or bastion of the general standards of public health, which start with the right to a safe home, are contravened.

To be fair to them both, planning permission and Building Regulations consent are 'statutory' matters, and their implementation is not of choice. As newer, better techniques in building methodology have developed over many years, and new strategies have been adopted in localities, it is little wonder that, even if they are better than what we had before, they morph into the 'rules', and become the newly raised bar that those in the know strive to beat.

The statutory control that applies to 'all' new buildings today is different to that which applies to older relics, in that not all older buildings and areas will be affected by the legislation that applies. To discover what is, and what is not, means a trip into the legislation, much of which has been amended since inception. A good place to start in England and Wales, however, is the 'Planning (Listed Buildings and Conservation Areas) Act 1990'. Its equivalents in Scotland, and Northern Ireland are 'Planning (Listed Buildings and Conservation Areas) (Scotland) Act 1997' and the 'Planning (NI) Order 1991'.

Although the legislation between the countries that make up the United Kingdom differs, the basics of what the legislation is trying to achieve in each country is broadly the same. Buildings that are thought to be 'special' typically fall under the legislation, and later in this chapter what are considered to be 'special' buildings are more particularly described. On the whole, however, these buildings will have architectural merit or will be of historical interest. Those buildings that fall into this category will be recorded on a 'list', hence the term 'listed'.

Once on a list, there are a whole host of matters that a building's owner will need to comply with, and included in the legislation is guidance on the authorisation of works that can and cannot be carried out at the property, with a whole range of penalties if the legislation is not complied with.

In every instance, the legislation in each country has been supplemented by various pieces of government guidance, all of which are long and detailed, and beyond the scope of an introductory chapter such as this. However, the essential facts are that 'listing' is concerned with the preservation of older buildings and areas of special interest are known as 'conservation areas'.

Listed buildings

A 'listed' building is one that has been entered into what is a list of buildings thought to be of 'special architectural or historic interest'. In the UK there are now upwards of 440,000 'listed' buildings, representing about 2 per cent of the latest total housing stock figure available of 22,564,000. Some of the entries do include multiple units within them, so the actual figure could be a little higher.

Not all of the properties that are 'listed' are considered equally as important as each other though, and are graded according to how important or special they are. *Grade I* applies to all buildings of exceptional interest only. Of all 'listed' buildings, this only applies to about 2 per cent or 8,800 in all.

*Grade II** refers to buildings that are thought to be particularly important and which are more than of special interest; 4 per cent of 'listed' buildings, or about 17,600, currently fall into this grade.

Grade II applies to buildings of special interest which warrant every effort being made to preserve them. The remainder, or 413,600, of all 'listed' homes are in this category.

The lists are open to public scrutiny and make fascinating reading for those that find historic buildings of any interest. A typical entry is given below; this is for a 'listed' home in Great Canfield in Essex:

> Hall house, medieval, extended in early C19, altered in late C19, timber framed, plastered, roofs partly tiled, partly slated. 3 bay block jettied to the E, early C19 extension further S. C19 external chimney stacks on N and S walls. 2 storeys. 3 late C19 bay windows, of which the middle one is below the jetty and late C19 gabled porch. First floor, 4 early C19 16 light sash windows. Roof of early C19 S extension of shallow pitch, hipped and slated, others gabled and tiled. Access refused in December 1982.
>
> In plan, this is a typical medieval hall house, but without internal inspection, it is impossible to know whether the walls of the hall have been raised to their present height, or whether the hall block has been rebuilt, as was common 1600 and later.

Clearly, the inspector was unable to get inside, and struggled to identify the 'real' property from beneath a cloak of newer extensions and additions. However, as was typical in previously unregulated days, many properties of historic interest were altered, and that regulations now exist to at least record what is left of our historic heritage should be applauded.

But regulations are nonetheless open to abuse. In the 1990s, an elderly gentleman took it upon himself to roam Essex, Suffolk and Norfolk to check upon all media that recorded properties for sale, and in those instances where older properties were shown or described that were not 'listed', but that he thought should be, he made an application to English Heritage for listing (it maintains the 'list'). You see, what is unique about 'listing' is that anybody at all who feels a property is of 'special architectural, or historic interest' can apply to have a building 'listed'; it need not be their own place of residence.

In many instances, as in the example given above, properties have been changed significantly over many years, but most building owners involved with instigating those changes are well aware of the need to enhance 'value', thus in most instances, those changes have been made sympathetically.

The gentleman in question was 'found out' when he made an application to list a property near Braintree in Essex. Outwardly, it displayed a wonderful plain peg tiled roof, all out of level; it had a range of old casement windows and displayed all the signs of an important house of distinction in a quiet backwater near the river Cam. The only signs of any recent changes to it had been a modern timber-clad garage to the right-hand side of the site and a 'granny annex' beyond the garage, used to house the occupant's elderly mother and father.

Once again, this clandestine, something-of-the-night chap appeared when the house was introduced onto the market for sale. He had fixed an appointment to view the property as a prospective purchaser. It is said by all those who met him that they first noticed him approaching the door furtively, precisely on time, and mostly after dark, hiding amongst the bushes best he could until he managed to gain access. He never took off his cap, which shielded his eyes; he must have been somewhat of a Walter Mitty type of individual, and he spoke slowly and deliberately, but never did he partake in unmeaningful conversation, and never was there any need to boil the kettle.

Occupants meeting him found that they soon became experts in bressumers, finials, jetties and the like, and any discussion about how nice a day it was seemed to be superfluous, and even discussing with him whether a coat and hat would be needed tomorrow was a no-go area.

Inspecting an old roof is a two part thing: outwardly, the appearance is majestic, the awkward cut of the tiles, the visible pegs are all to die for, the uneven ridge is a must, and all the signs that it is about to blow away in the wind, are essential for all later-day roof spotters. Just like members of the Eddie Stobart club who tick off Aunt Bessie or Florence Nightingale on their list as they purr past them on the motorway, or those with binoculars at the edge of an airport runway who drool as a Boeing 777 passes by just a few metres overhead, those who like historic buildings have a real 'need' to see the 'undercarriage' – they have to gain access into the roof space.

And it is important to add that although most internal alterations to a 'listed' building require permission under the guise of 'Listed Buildings Consent', many do not. If it is currently in your 'zone of contemplation' to install a loft access ladder, done properly it's unlikely to be worth an application to English Heritage, unless of course you are about to destroy a perfectly good historic artefact in the process.

Back to that property in Essex, yes, the gentleman did need to see the loft space, and a loft ladder was hurriedly sourced from the modern garage. He entered, ruminated, and cogitated at some length, and in a disappointed gruff voice, apparently said 'All show, no substance', and off he went with little further ado. That house you see had previously been thatched, and then slated, and only when the current vendor wanted to 'restore' some of its former glory had an 'old effect' roof been put on, the old roof having been changed about a year previously. He apparently left in a huff but, interestingly, missed some really old features: a lovely original 'inglenook' fireplace, a priest hole and, oh yes, a ghost who regularly made things colder than they ought to have been at the bottom of the stairs!

But even then, the owners were left in a dilemma: they were told by the 'listed buildings' inspector who attended as a result of the listed application being made by this gentleman that the property was borderline, it either could be 'listed' or not, the decision could go either way. Suddenly, they found themselves in an awkward position: would 'listed' status add or detract from value?

According to the latest edition of the Royal Institution of Chartered Surveyors (RICS) Professional Standards (sometimes known as the Red Book), the definition of value is:

> The estimated amount for which a property is expected to exchange after the date of valuation, and specified by the valuer, between a willing buyer, and willing seller, in an arm's length transaction, after proper marketing, wherein the parties had both acted prudently, knowledgably, and without compulsion.

The 'Red Book' goes on to add commentary about what it thinks are relevant factors in the assessment of value:

> The apparent general state of/ and liability for repair, the construction, and apparent major defects, liability to flooding and other risks.

Going on to consider the 'nature of the interest', the Red Book adds that the valuer needs to be cognisant of 'Any restrictions, covenants, and other obligations . . . .'

So, there's the dilemma: any valuer would need to take into account that a 'listed' property brings with it certain restrictions on what you can do to it. However, in general, the key to the value of a 'listed' home is all about personal preference. Some potential buyers may want to proceed to purchase with great haste, because they want to live in a historic home. They would say that their purchase was 'without compulsion' as a result. Others might shy away on 'knowledge' or 'prudency' grounds, given that effecting repairs might be a little more costly than for a standard home, or that running costs for a poorly insulated old relic would be higher than that of a recently built home, fully compliant with current Building Regulations.

In the case of that property in Essex referred to above, given the choice to list or not, the owners decided not to proceed with 'listing' – their property was very pretty in any case and would sell easily whether it be 'listed' or otherwise, and the 'choice' baton could be passed on to other future custodians of the house. Interestingly, the property did sell to the first genuine viewers some two days after being viewed by that rather shady character, and completion took place only some six weeks thereafter.

However, not all 'listing' is as hit and miss as this and, as a general guide, buildings are chosen for a variety of reasons, but the older the property is, the more likely it is to be included in the register. Buildings that were built before 1700 and which have survived to a large extent in their original condition, for example if they still have original features worth noting, are almost always 'listed'. Buildings built between 1700 and 1840 that are of good quality and have some character are generally included in the register as well.

After 1840, the boom in building really took off and a considerable number of properties have been constructed since then, some with merit and others without. However, being relatively young, vast numbers of them have survived to this day. In this category, therefore, only the very best examples of quality and character are even considered suitable for 'listing'.

After about 1914, it is fair to say that buildings have to be of particularly outstanding quality or 'iconic' in order to be included in the register. Today, no buildings less than 10 years of age are 'listed'.

Interestingly, not all 'listed' buildings fall within the conventional description of a home. English Heritage will consider any 'building' which has special architectural or historic merit, and that might be one that is not for the living, but for the dead. A good example of this exists at School Green, East Leake in Nottinghamshire. A 1914–1918 'war memorial' is listed Grade II (see Figure 2.1). The 'listing' text states:

> War memorial, circa 1919. Ashlar, some copper. Hexagonal platform topped with chamfered octagonal base. Over is a hexagonal column with moulded base and scalloped capital surmounted by a single copper cross.

Everyone or almost everyone thinks of 'listed' buildings as being old. Indeed, this chapter has already on a number of previous occasions referred to them as being 'old relics'. However, this is not always the case.

Figure 2.2 shows the iconic Willis Corroon building in Ipswich, Suffolk, UK. It became famous in 1991 when it became the youngest building ever to have been considered by English Heritage for statutory protection, and it was 'listed' as Grade 1.

Now after intimating that the majority of 'listed' buildings are old, here it is in all its glory, a magnificent 'old' relic. Hang on, re-wind, just take a look at what's in front of you: it's all

Figure 2.1 War memorial, East Leake

Figure 2.2 Willis Corroon building, Ipswich

dark glass, and a funny shape. What's to like in that, and aren't there are, after all, many similar ones in the UK and the world over?

Yes, there are, but this building was one of the first commissioned from an emerging, great architect – Sir Norman Foster – famed for many buildings including London's Stansted Airport, the 'Gherkin' building in London, and the new Wembley Stadium, home to the England football team. Work to build it started in 1970 and ended in 1975; it was one of the first buildings to be built in a style known as 'high tech'.

The 'high tech' parts of the Willis Corroon building are many, but particularly state of the art for the time are its grid of concrete pillars which hold everything up, fourteen metres apart, supporting cantilevered concrete floor slabs. It was impressive for its time, and indeed still is today, a forerunner for many buildings which have emerged since.

It is noteworthy that many iconic buildings have the same designer: Sir Norman Foster's business, Foster + Partners, boasts nearly 200 awards for design excellence, and it has won over 50 national and international competitions. Overseas excellence has been recognised for his designs of the new Hong Kong International Airport built between 1992 and 1998, the Congress Centre in Valencia, 1992–1998, and the new German Parliament Building, the Reichstag, 1992–1999.

Conservation areas

English Heritage is responsible for policing areas which have been designated for their special architectural and historic interest in London. However, it is usually a local authority which has the responsibility for 'designation' within their own area.

Nowadays there are over 8,000 conservation areas in the UK, and those living within them will find a whole set of rules to comply with in the event that they want to make changes to their properties. Those either lucky enough, or unlucky enough, to live in one of these areas are likely to need to obtain permission for any alterations that may impact on the visual aspects of the area. Typically, these will include changes to the appearance of the property by adding or removing cladding, changing the windows, the building of boundary walls, laying of pathways, changing the roof covering, etc.

The requirements are also likely to impact on television viewing habits as satellite dish positions will also be subject to the rules. Never mind if the only viable position for the dish is at the front of the property, in the event that the dish detracts from the look and feel of the area, it is unlikely to be permitted, whether or not Walthamstow Avenue are playing away at Manchester United.

The basis of the idea is to preserve historic areas of interest for the majority, and although it is often heard said by the minority that the rules are an imposition, there is a certain 'nimby'(ism) about those enjoying lovely areas, that understandably others would like to enjoy too. But it can be a pain. Finchingfield in Essex is about 8 km north of the pretty commuter village of Great Dunmow, which itself is about 5 km from Stansted Airport. Great Dunmow is famous for its 'Flitch' celebrations.

The 'Dunmow Flitch Trials' are held every four years, and they award a flitch of bacon to married couples from anywhere in the world if they can satisfy the judge and jury of 6 maidens and 6 bachelors that in 'twelvemonth and a day', they have 'not wisht themselves unmarried again'.

A 'Flitch' is a side of bacon, and it is ceremonially paraded into the centre of Great Dunmow during the trials. There is a reference to 'The Dunmow Flitch' in *The Wife of Bath's Tale* in Chaucer's fourteenth-century *Canterbury Tales*. So every fourth year Great Dunmow becomes a hive of activity as hundreds of people descend upon it and enjoy all that a pretty Essex market village has to offer. At the end of proceedings, they all go home, and normality is restored to this usually sleepy area.

Residents in and around the Finchingfield locality, however, do not consider themselves quite so lucky. The village has become a common tourist coach destination because it is generally regarded as being very 'pretty'. In fact it has won competitions. Views of the well-known green, pond, cottages and church are often to be found on calendars, chocolate boxes, tea towels, postcards, jigsaws and the like. It has been described as 'the most photographed village in England'.

Figure 2.3 Great Dunmow

So, every summer, when the weather turns for the better, hordes of coaches head to the village at weekends and disgorge over 50 passengers at a time to enjoy its undoubted beauty. Those coaches travel through the Essex countryside, up and down single-track roads meant for horses and carts, and sometimes disturbing horses out for an afternoon stroll with their riders. The coaches drive to undesignated parking areas where they have to fight for space with the local motorbike run, thirty or more bikers who have driven their own petrol guzzling horses sometimes over a hundred miles through plenty of supposedly undamaged countryside, and all to enjoy an hour of solace in the sun, enjoying a previously unspoilt area, which is now gradually beginning to feel the strain and deteriorate. This scene is played out across the whole of the UK in nice places, every summer.

And, of course, as the late afternoon sun begins to set, off they all go: diesel coaches, and petrol bikes, all horses of a type, laden with those that have enjoyed their cream teas in the Finchingfield Tea Rooms or a beer at the Red Lion Inn. The well-fed ducks settle back into the pond, and normality returns for another week.

It is an interesting debating point whether in an effort to preserve an area, that measure itself in fact causes damage, or at least creates the conditions for damage to occur.

There are very many reasons why we like to visit places: to some it is to see the beauty that unfolds during the journey to get there, and to others, it's a chance perhaps to re-live the places that we used to visit as a child. It may of course just be going somewhere nice: we are bound to take something away with us, whether it be a small piece of the grass verge as the coach has to pull over to let a cyclist going the other way pass by, or just a new memory, or a memory renewed, but just as black is to white, night is to day, a small part of our destination gets damaged, and 'that' is why preservation and protection measures are at least tried by those charged with saving our heritage.

So, there's the rub. 'Not to protect', and hope nobody visits, so there are no changes to the appearance of a particularly nice area, or 'protect', which tends to stimulate interest; and watch protected areas suffer under the strain. We do after all protect areas that we consider

Figure 2.4 Finchingfield

to be 'Sites of Special Scientific Interest' (SSSI's); these are conservation designations that protect a particular area, they are the building block of what has become the 'nature conservation' legislation in the UK. However, such legislation is unlikely to be rolled out to protect areas that are just historic, and 'pretty', rather than essential to conserve. Long live the protected Great Crested Newt.

Scheduled ancient monuments

There are a wide variety of archaeological sites, monuments and structures which range from standing stones to Second World War pill boxes. These can also be protected as ancient monuments worth keeping.

There is, of course, a difference between these and 'listed' buildings, in that scheduled monuments do not generally have a day-to-day viable use. The relevant legislation is the Ancient Monuments and Archaeological Areas Act 1979. In short, all works to these monuments and in these areas require what is called 'scheduled monuments consent'. Any work undertaken to them without this can lead to a criminal conviction and a fine. Interestingly, in these days of metal detectors, and hordes of people scanning fields for treasure, under this Act it is also a criminal offence to enter these areas or go to one of these monuments and use a metal detector, and even worse if anything found of value is then removed.

At the moment, there are about 30,000 sites protected by over nearly 19,000 entries in the UK 'schedules'. In England, they are overseen by English Heritage, in Scotland by Historic Scotland, and in Wales by Cadw.

Force Crag Mine (Figure 2.5) is an example of a scheduled monument: not everyone's idea of an historic monument but an important survivor from an industrial past, and now the subject of careful conservation by English Heritage.

Figure 2.5 Force Crag Mine

Perhaps what underpins preservation legislation in the UK is not only maintaining the nation's 'heritage' but, for us, it's about maintaining our memories. In Cardiff, South Wales, is the St Fagans National History Museum. It is billed as one of Europe's largest open-air museums and Wales's most popular heritage attraction.

The site includes over forty of Wales's historic buildings and includes a 'Celtic Village' comprising three circular houses, based on originals in Gwynedd, Worcestershire and Flintshire; on many days during the year, actors re-create life in a typical Celtic village, by using Celtic tools, and crafts of the day.

There's also St Teilo's Church, built originally in stages between 1100 and 1520, and moved stone by stone to the museum in recent years. A property known as 'Kennixton', a good example of a typical Welsh farmhouse from the Gower Peninsular in South Wales, is also to be found here. As you walk around the inside, its blood-red walls are immediately visible, this was originally thought to keep away evil spirits, and many properties of its day were painted in this colour. Together with the age-old technique of carving into the front entrance hallway the image of berries on a Rowan tree and carved figures, no evil spirit dare enter this home.

Then there is an exhibit called 'Rhyd-y-Car', a terrace of ironworker's cottages built originally in 1805. They house furniture and objects that illustrate the different periods in the history of Merthyr Tydfil, a Welsh mining community, up to 1985.

The 'Workmen's Institute', formerly used by all manner of groups and societies, is one type of building which typifies many others around the UK, ranged under the banner of buildings used as part of the 'Miner's Institutes' and 'Workmen's Halls'. It is an original building, moved to the site from its former position a short distance away.

There are many other historic buildings on the museum site, including an old post office, an old school, still equipped with its desks and inkwells, an old chapel house, and many other traditional buildings.

Many people enjoy a good day out at St Fagans, particularly if the weather is good! Ice creams intermingle with the roaring log fires, and often visitors are heard to ask why in one of the historic, re-located farmer's cottages there's hay on the floor. 'Ah!' starts up the very willing host of that particular house. 'See that bit of wood under the door? These hay's are the "threshes", and that bit of wood is the "threshold". When the hay, which is put down for comfort and cleanliness, reaches the "threshold", that's when it gets removed. That is the origin of "stepping over the threshold", when entering an old historic house.'

Good things are also happening to encourage communities to become involved in the look and feel of their own areas. In order to help rural communities maintain their character and manage change without altering the uniqueness of their area, the government has now established the concept of 'Village Design Statements' (VDS). VDSs give a detailed description of the existing character and main features that exist in an area, and the things that make it special for the residents. Local parish councils have been charged with their preparation and those that have so far been published show an overwhelming desire to maintain the tranquil, rural characters of village.

VDSs are prepared within the context of the local authority's adopted local plans to ensure that their stated policies and guidance are complied with. Subsequent to their preparation, and adoption, the VDS becomes part of the planning process for the area.

Most VDS provide guidance for the local community. One such is the Great Canfield VDS which perhaps summarises best what we as custodians of our own properties, the areas in which we live and, in fact, the UK as a whole can do. It says:

All alterations affect the building, and its surroundings, as well as the overall look of the village, so please make your own assessment of the potential impact by studying each visible elevation of your property, including the rear elevation and ask yourself; what are the distinctive features of your property, neighbouring properties, and the area? How do the alterations you are considering affect the positive distinctive features of your property? Do they complement the character of the local area? Do they meet the guidelines set out in the VDS? If not, how could you change them so that they do?

The VDS then goes on to add: 'Additionally, is there the opportunity to remove any uncharacteristic features?'

In short, it encourages us all not just to consider what is best for us, but also to consider others; after all, that is what the designer of our 'listed' homes would have liked, whoever they are, and if their work can be shown off to everyone, in undamaged, pleasant areas, then all the better. That after all is what underlies 'preservation', and 'conservation' and makes our homes and our communities better places to be and, above all, enhances our lives.

3   Sustainability

There is a rather neat definition of what is 'sustainability' available to those who search for just that question on the internet. Those engaged in the debate suggest that it is about 'meeting the needs of the present generation, without compromising the ability of future generations to meet their needs'. How that can be achieved often varies, but there is a general agreement amongst those who are 'in the know' that it involves questions about how we reduce the consumption of the Earth's dwindling resources, how we can all become more fuel efficient, and how we can recycle even more.

It is suggested by some that the very word 'sustainability', and all the statistics that follow it, has become the umbrella word, perhaps the 'watch' word for measuring the outcome of how we, as humans, act whilst we are on this Earth.

If certain statistics show that the Earth is more poorly tomorrow than it is today, then our collective actions have become less sustainable in the longer term. In the event that there is an improvement, then we have done well, and have started to move towards saving the Earth.

For those of us perhaps not as engaged in the debate about sustainability as some, luckily there are 'those in the know' who produce guidance for us to read as well as legislation. Guidance gives us useful advice, whilst legislation potentially saves us from ourselves!

The primary legislation which drives 'climate change' thinking in the European Union is called the Energy Performance of Buildings Directive' (EPBD). It is designed to alert us as to how buildings perform in relation to energy, and requires the production of different types of 'Energy Performance Certificates' for both privately owned and public buildings either periodically, or when sold or let; there's also a requirement to inspect air-conditioning systems and certain size boilers from time to time. In the UK, there is strong governmental support for the Directive and for current proposals to 'strengthen' the scope and application of its requirements. These will, it seems, be a vital tool in the box if the UK is to meet its declared target of reducing carbon emissions by up to 80 per cent by 2050. This, on the face of it, appears sensible if one accepts that energy used in buildings in the UK accounts for up to 50 per cent of all carbon emissions in the UK.

Tagging a building with information about its *energy performance* is intended to spur building owners into making energy-saving improvements. Where, for example, there is no insulation in the loft, a better 'energy performance' certificate result will be had if insulation is added. Insulating cavities in external walls, double glazing windows, weather-proofing against drafts, etc. are all easy to achieve measures that will improve ratings. 'Energy Performance Certificates', provided by accredited energy assessors, grade the energy performance of a building using letters A to G. An A grade is awarded to the most energy efficient properties, whilst a G grade is awarded to the worst performers. In the UK, the average grade awarded is D; this suggests that the bulk of the UK's property stock is very inefficient energy wise, and that means it is more costly to run.

More importantly, however, for those engaged in the 'climate change' debate, the latest data available from the World Bank's 'World Development Indicators' suggest that, per capita, the UK emits 8.84 metric tonnes of carbon into the atmosphere every year. In fact, there has been a steady improvement in the UK's carbon emissions statistics over the past few years: it peaked in the early 1970s at around 12 metric tonnes per capita while in 1990 the UK emitted around 10 metric tonnes per capita.

However, there is still a significant amount of work to do around the world to lower the total amount of carbon emissions. China is developing rapidley: in 1960, China emitted about 1.5 tonnes of carbon per capita; by 2007, this had risen to around 5 metric tonnes, still well below UK, but on something of an upward trend. Russia, conversely, has seen a significant improvement in recent years, down from a peak of around 17 metric tonnes per capita in the late 1980s to around 10.5 tonnes today. The United States' emissions have been fairly consistent since around 1983; the USA now emits about 20 tonnes of carbon per capita per annum according to latest World Bank figures. There is now a general acceptance worldwide that there is a need to cut carbon emissions, but the steadily increasing amount of emissions from some countries is of concern. For example, if the current upward trend is not reversed, by 2030 China will be close to emitting nearly 12,000 million metric tonnes of carbon into the atmosphere every year.

'Climate change' is now accepted as being a generic term that covers long-term changes in average weather conditions, including temperature and wind. The United Nations Inter-governmental Panel on Climate Change (IPCC) is made up of many of the world's lead-ing scientists, who are experts in the field of climate change. According to the IPCC, the world's climate has undergone dramatic changes in the past few years. They say that this is the direct result of the effect of 'greenhouse gases'. Their statistics show that as a result of human activity, not the least of which is carbon emissions, average temperatures around the world have risen and, as a result, there are rising sea levels and ice melts. Worldwide average surface temperatures during the twentieth century increased on average by 0.74 degrees Celsius. Worryingly, the IPCC forecasts that this trend will continue, and indeed accelerate during the twenty-first century. As a direct result, they forecast that there will be a detrimental effect on the weather with, for example, an increased number of hurricanes, and that this will be exacerbated towards the end of the century, by which time global sur-face temperatures will have increased by a further 1.1 degrees Celsius. Unless we improve our emissions performance, this spiral of decline is set to continue.

Greenhouse gases are said to be the main culprit. They are an accumulation of gases that hang around in the atmosphere, much like a blanket. Others describe them as a type of glass ceiling or domed top covering the Earth. The importance of this is that escaping heat that would ordinarily be able to dissipate is unable to do so, and the Earth warms up as a result. This is called the 'Greenhouse effect'.

On the other hand, there are those that do not believe that climate change is real. Since 1995, the IPCC has conducted four detailed assessments of climate change, and on each occasion it has reported, it seems to be with a greater amount of confidence that there is a case for the climate being changed by what it calls human induced activities. In 2007, it concluded, 'it is extremely unlikely that the global climate changes in the past 50 years can be explained without invoking human activities'. Other potential causes such as a change in the output of the sun and the effect of volcanoes have been considered, but there is now widespread support in the scientific community that change in human emissions behaviour is needed, and that this change must be effected quickly.

Some consider that simple improvements in the home, to our flats and houses, or to our factories, schools, public buildings and offices, will have little long-term effect on climate change. It is true to say that they will have no impact outside of the UK, but if the UK can reduce its CO_2 emissions per annum per capita, and that downward trend is replicated across the world, the IPCC remain confident that the greenhouse effect can be reversed. It is, however, a familiar trait in the UK that little change is effected without the promise of pecuniary advantage. But in an era in which the UK and the global economy as a whole is trying to emerge from the devastating effects of the global banking crisis, which started in 2007, there is precious little, if any, money available to make improvements to our property. Indeed, the 'credit crunch' was a devastating blow to many around the world, and may eventually be seen as being one of the first nails in the coffin of the Earth.

Of course, there is a direct link between home improvement expenditure and savings, most manufacturers of home improvement products shout from the rafters about how much in energy bills will be saved by even the most modest home improvement, but in a time of a real need to save money, even the cheapest roll of insulation, and even the thinnest and cheapest therefore of double glazed inserts, may be seen as unnecessary expenditure, and nothing gets done. The most recent *English Housing Survey*, dated 2008, and known as the '*Housing Stock Report*', was issued before the full impact of the 'credit crunch' was felt in the UK. It reported that the energy efficiency rating of UK homes had steadily improved since 1996, at which time the average SAP (Standard Assessment Procedure) points per UK home stood at 42. By 2008 the figure was 51. Nonetheless, the survey highlighted that 20.1 million dwellings could potentially benefit from one or more cost-effective improvement measures such as loft and/or cavity insulation, upgraded boilers, and better heating controls. They concluded that if such improvements were carried out to the average dwelling, the SAP rating would improve by a further 11 points to 62, and CO_2 emissions per property would fall by 26 per cent. However, no available money has meant that in 2009, UK expenditure on DIY (Do It Yourself) fell by 4.5 per cent, the result being that the UK home improvement sector returned to 2004 expenditure levels. This is not good news for CO_2 emissions per capita, and it is not good news for the Earth.

If money had been available to continue the trend of improvement in SAP ratings, it is likely to have been used to improve a number of obvious heat loss areas. The first of these to address is loft insulation. According to the Energy Savings Trust, most heat is lost through a lack of proper insulation in a building, and a high contributor to this is the lack of insulation in the roof space, either situated between the ceiling joists, or between the rafters. The recommended thickness for domestic homes is 270 mm of mineral wool.

Heat will always pass from a warm area to a cold one. So, the colder the area is outside a property, the keener warm air is to escape. By creating a barrier, it is more difficult for the warm air to escape, particularly if the insulator, such as mineral wool, contains air pockets within it to trap the warmer air. The effect of this is to reduce what is called the 'U value' of the property. A 'U value' is an indicator of how quickly a property loses heat. Thus, the lower the U value, the more energy efficient is the property. The Energy Savings Trust suggests that an uninsulated roof has a U value of around 2.3, and that this falls to around 0.16 once 270 mm of mineral wall has been introduced. This is an improvement of some 95 per cent, and cannot help but reduce energy costs. Table 3.1 shows figures from the Energy Savings Trust detailing how much per annum an insulated loft space saves.

Figure 3.1 Loft roll insulation

Table 3.1 Loft insulation

	Increase in loft insulation thickness (0–270mm)	Increase in loft insulation thickness (50–270mm)
Annual saving per year (£)	Around £145	Around £40
Installed cost (£)	Around £250	Around £250
Installed payback	Around 2 year	Around 6 years
DIY cost	£50–£350	£50–£350
DIY payback	Up to 3 years	1 to 9 years
CO_2 saving per year	Around 730 kg	Around 210 kg

These are estimated figures based on insulating a gas-heated, semi-detached home with three bedrooms. The mineral wool loft insulation available today is expected to be effective for at least 40 years, and it will pay for itself over and over again in that time. The better insulated the home, the less energy will be needed to keep it warm, thus the more money that will be saved in the long run.

By saving energy, a household will produce less CO_2. So, adding or topping up loft insulation is a great way to start reducing the impact of carbon emissions on the environment. If everyone in the UK installed 270 mm loft insulation, the energy savings would be around £520 million per annum and nearly three million tonnes of CO_2 every year. That's enough to fill the new Wembley Stadium nearly 380 times.

Double glazing, mostly because of over-zealous double glazing sales people, has developed a bad name. Over-priced and poorly installed windows and doors, financed short-term by high annual percentage rates (APRs), has meant that many of those people that might have installed new windows and doors have shied away. Since 1 April 2002, all replacement installations have had either to have achieved a successful Building Regulations application prior to commencement of the work, or alternatively, the installer must be a member of the FENSA scheme. The principal driver for the FENSA scheme is stated to be a desire to

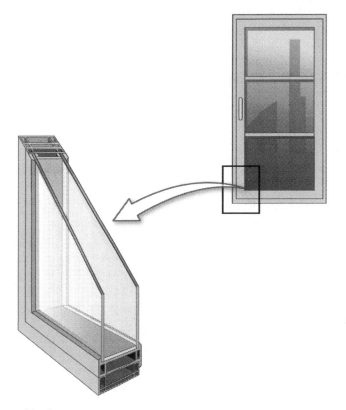

Figure 3.2 Double glazing

ensure that only products that perform to a particular energy efficiency standard in respect of heat loss are installed. The FENSA scheme was set up by the Glass and Glazing Federation (GGF) at the request of the Department for Communities and Local Government (DCLG). FENSA stands for Fenestration Self-Assessment Scheme.

Fifteen per cent of all heat loss is thought to be through cold or draughty windows. According to a guide produced by the London Borough of Waltham Forest, a typical £4,000 capital expenditure on Building Regulations compliant double glazing will save around £135 per annum, dependant of course on the number of windows installed. At these rates, the typical payback period before real savings are made is around 30 years, longer if the project is financed and interest is payable. Different levels of expenditure will clearly impact on this period but, in short, the installation of double glazing is not a scheme that will pay dividends in respect of costs versus energy savings quickly.

Double glazing acts by creating a vacuum between two pains of glass, and heat loss is reduced. Additional benefits include better sound insulation and potentially, less condensation.

However, if the payback period is very long, those thinking about such home improvements might not have the inclination to do anything at all about their property wasting energy if there is nothing really in it for them. The key in such instances must be 'state grant aid' – or is it?

The latest advice from the Department for Business Innovation and Skills (BIS) is that 'state aid' is now called 'regional aid'. 'Regional Aid' can be used to promote the economic development of certain disadvantaged areas within the European Union, under Article 87(3)a, and Article 87(3)c of its 'Treaty'. In the UK, the main forms of state aid are through 'discretionary grant schemes'. In Great Britain and Ireland there is a scheme known as 'Grant for Business Investment' (GBI), which is administered by the BSI and also, until recently, by the Regional Development Agencies (RDAs). The main thrust of this funding is through investment in projects that lead to long-term improvements in productivity, skills and employment.

In Scotland, there is 'Regional Selective Assistance' (RSA), which is administered by Scottish Enterprise and is aimed at encouraging new investment projects, strengthening existing employment and creating new jobs.

In Wales, the scheme is called the 'Single Investment Fund' (SIF), which is delivered by the Welsh Assembly Government, and is there to help support new commercially viable capital investment projects that create, safeguard and maintain permanent jobs.

This approach seeks to support the infrastructures of business itself, rather than making of money available to undertake energy-saving work. However, more targeted help is available through a number of sources. For example, there are a wide range of 'Disabled Grant Facilities' currently available to those who need to make changes to their homes, and these potentially include changes that impact on energy saving.

Local authorities also possess certain discretionary powers to provide help towards repairs, improvements and adoptions, and for works such as these they are the starting point for the initial enquiry. They can also advise about other 'home improvement' agencies, often called 'Care and Repair' or 'Staying Put' agencies, that will give useful advice on such matters as how to adapt, repair, improve and maintain a home. This assistance is usually focused on giving advice on what welfare benefits are available to help.

Of course, improving ones home is not just about home insulation, or double glazing, there are many other changes and improvements that can help too. The use of 'energy efficient appliances' is often something that is overlooked. In the UK, refrigeration of foods, etc., is an area that we spend around £1.5 billion on every year – on buying the appliance, and on fuel costs running it when installed. Fridges and freezers are on 24 hours every day, seven days a week, 365 days a year, so using energy-efficient ones is likely to save us money. More than 90 per cent of all homes in the UK have a washing machine; about 40 per cent have a tumble dryer. Washing machines and tumble dryers use hot water, so once again using the more energy-efficient models can save us more money. The money we save on using energy-efficient appliances can then be used wisely around the home, perhaps to save even more energy, like buying and installing 'energy-saving bulbs'.

It is thought that if we replace a conventional 100 kWh bulb with energy-saving bulbs of the same specification, we will save £9 per bulb per year (and 50 kg of CO_2 emissions). This does not, on the face of it, sound very much, but in a typical 1930s semi-detached house there will be three bedrooms and a bathroom upstairs, all with at least one bulb, two reception rooms, a kitchen and possibly a downstairs loo, all needing light, and then there are the entrance hallway and upstairs landing – that's ten bulbs in all, or a saving of £90 per annum. Many local authorities give these bulbs out free of charge, the savings can well worth the effort.

What about installing an energy-efficient central heating boiler? Changing a boiler is a costly event; rarely will a boiler cost less than £1,000 to supply and install, and often much more. Most exiting boilers in the UK are thought to be pre-1990 stock. They will not, therefore, be efficient and are unlikely to be condensing boilers. That means that they are likely to be of a type that contributes to up to 60 per cent of all domestic carbon dioxide emissions. The savings attributed to efficient boilers does vary and depends mainly on the model, but, 'on average', an energy-efficient boiler will save around £250 per annum, thus paying for itself in 5–10 years.

Older boilers still to be in use in the UK are likely to have a hot water cylinder as part of the heating system. Fitting a 'British Standard' jacket to a domestic uninsulated cylinder cuts heat losses from the cylinder by up to 75 per cent. A typical insulating jacket will cost around £20. The potential savings will be around 550 kWh, or 110 kg CO_2 per year, which at current prices is about £25–40 per annum. So, a new jacket will usually pay for itself within a year.

Other domestic improvements that will cut CO_2 emissions include adding insulation to the external walls of a property. Often this is only considered possible where a cavity gap exists, but this is not the case. Solid walls lose heat more quickly than cavity walls. The majority of pre-1930 and also some post-1930 properties are usually of solid external walls construction. Solid walls can be insulated by applying insulation either to the internal face of the external walls, that is within the home, or as an alternative, attaching it to the outside, then covering it over with a suitable weather-proofing material. If it is to be fixed to the external face, then additional considerations will have to include things such as the impact it will have on appearance, but these days there are a whole host of alternative types of finish available ranging from modern claddings to coloured renders which can achieve an attractive appearance.

If the insulation is to be placed internally, once again there are things to consider, such as that there will be a small amount of loss of internal floor space. Again, a decorative finish will usually be applied over the insulation, like a simple render, with either wallpaper or a painted finish. Whether the insulation is fitted inside or outside, it usually consists of mineral fibre or an insulating board, both of which typically are about 50 mm in thickness.

For insulating cavities, there are various methods which include injecting insulation either externally or internally, but with this type of construction, the key is to ensure that no damage is caused to the cavity ties which hold the inner and outer leaves of the wall together, either during the work, or by any chemical reaction thereafter. According to the 'Energy Savings Trust', the cost of insulating external walls will vary, but 'on average' the energy saving will be around £110 per year. In terms of carbon emissions, the average saving is thought to be around 1,600 kg CO_2 per annum.

Of course, there will be homes where achieving any sort of energy saving is difficult. The 'Energy Savings Trust' calls these types of home 'hard to treat'. They say that there may be around 10 million UK homes for which conventional energy-saving measures will not work, like flats in high-rise tower blocks, rural homes built of boulders and stone, those with solid uninsulated flat roofs, etc. In such instances, it is suggested that the use of 'renewable energies' might be useful.

The best definition of 'renewal energy' is an energy which comes from natural resources such as from sunlight, wind, rain, tides and geo-thermal heat. The term renewable means 'naturally replenished'. One of the most noticeable changes to the landscape in the UK over the past few years has been the emergence of wind turbines both on and off shore. The use of wind power in the UK is said to be growing at a rate of around 30 per cent annually. According to Wikipedia, the UK currently has a capacity of about 158 gigawatts

Figure 3.3 Wall insulation

(GW). In the past couple of years, improvements in the technology of photovoltaic cell (PV) installations has seen an increased uptake, and the older solar panel technology still remains in use.

One of the attractive features of renewable energies is that whilst there are many large-scale projects, their technologies are suitable for use in domestic homes. Globally, it is thought that around 3 million households already get power from what are called 'small solar PV systems'. Also, there are 'micro-hydro' systems that can be configured as mini grids serving village and rural communities, and it is thought that 'biogas' made in single-household-sized anaerobic digesters is used in around 30 million rural households for lighting and cooking. 'Biomass' cooking stoves, which use waste products from organic material such as wood, and waste gas, are thought world-wide to be in use by about 160 million households.

As mentioned earlier, in the UK there is a tendency not to enter into significant expenditure if there is 'nothing in it for me'. However, there is something on the horizon that might force a change if the motivation is money. Above, we briefly mentioned 'Energy Performance Certificates'. Remember, an energy efficient home might be awarded the accolade of an A rating, whilst one that allows almost all energy to escape will receive a paltry G grade. Given that the average UK home gets a D grade, there are murmurs amongst those that set our legislation (although they would deny it) that 'something must be done'.

Imagine two identically built homes, perhaps somewhere in suburbia, a pair of semi-detached houses, built with a simple pitched and tiled roof and cavity walls in the 1960s. The one on the left has loft insulation, double glazing, cavity wall insulation, a condensing heating boiler, energy efficient bulbs, etc., etc. It has cost a fortune to achieve all this, but the owner has just been awarded an A grade by an 'energy assessor' who had been employed to visit the home. The neighbour on the right, who is less fastidious, has used all his money growing vegetables in the garden; however, whilst he has been able to enjoy fresh carrots, swedes, and an endless supply of potatoes, although he feels great, he has allowed his home to fall into disrepair. There's no insulation in the home at all, the windows and doors are draughty, and the only way he can keep warm is to use an old three-bar electric heater; he can't use the fan heater because that's in use in the greenhouse warming the next batch of organically grown tomatoes. He also wears an endless supply of old donated coats. If that same 'energy assessor' had visited his home, it surely would have been awarded a G grade.

At present, both properties having an identical layout and being in the same location would have been rated for Council Tax in the same way. Council Tax is a tax levied on households by local authorities; it is based upon the estimated value of a property. The relevance of 'value', however, and differences in condition is diminished when one knows that the Council Tax bands are widely drawn. This tax was levied after the introduction of the Local Government Act 1992 and replaced the discredited 'Community Charge', known as the 'Poll Tax', which was to be shortly introduced.

Council Tax has eight 'bands' and every residential property in the UK has been placed within one of them. Band A is up to a value of £40,000. Bands B to H increment as follows: £12,000, £16,000, £20,000, £40,000, £40,000, £160,000. The last band, which is band H, applies to all properties worth over £320,000. Thus, there is a wide value difference in each band, and properties that might be very different in condition can often find themselves paying the same level of Council Tax.

Returning to the earlier scenario, the healthy vegetable eater will potentially be paying the same Council Tax as his neighbour, despite his property being in a poorer state of repair. The rationale behind this is said to be that the tax should not benefit those that allow their properties to deteriorate in condition. However, this is a negative approach and the balance of benefit for looking after one's home is starting to turn.

If the UK is to meet its stringent carbon emissions reduction target over the next few years, it is going to have to be innovative and tackle the 'what's in it for me' complacency that abounds in households across the UK. Why not therefore calculate the Council Tax payable according to one's carbon emissions from the home? In short, the better the 'Energy Performance Certificate' (EPC) grade, the less you pay; the worse the grade, the higher the tax.

There are, of course, stumbling blocks to all this. In 1990, shortly before the introduction of the 'Poll Tax', which was to replace the old 'domestic rates' system, the idea of placing a responsibility on each individual occupant in a home to pay a proportion of the tax due did not appeal to everyone, and a spate of riots occurred across the UK. By far the largest of the disturbances took place in Central London on Saturday 31 March 1990, with rioting and looting of shops and offices starting at 11 am and not finishing until around 3 am the following morning.

So, a tax on the energy efficiency of people's homes might prove divisive, given that the benefits might only be available to those who have sufficient funds available to improve their homes, and although the general idea of improving the energy efficiency of the UK's housing stock would prove attractive, the impact of its introduction would need to be very carefully considereded. Thus, the UK's target of reducing our carbon emissions by 80 per cent by 2050 might turn into an unwelcome battle. A shake up in grant funding available is likely also to be necessary, but if this is to be successfully achieved, a sea change in people's thinking will also be required.

In 2009, the US consultancy firm McKinsey issued a report entitled 'Energy Efficiency is the Answer'. Strategic climate-change thinkers had been concerned with the failure of the US to reduce per capita carbon emissions year on year, and the McKinsey report looked at how greenhouse gases might be reduced. The main conclusions of the report were that, in the US, a reduction of about 1.1 gigatonnes of greenhouse gas emissions might be possible if some basic energy-saving measures were adopted by US households, including fixing 'leaky windows'. That, the report goes on to say, is the equivalent of taking the entire US fleet of light trucks and cars off the road, and an energy saving of around $1.2 trillion, if businesses and individuals spent about $520 million on energy-saving initiatives.

Reading between the lines, its comment 'This dispersion ensures that efficiency is the highest priority for virtually no-one', could well be applied to the views of the general public in the UK. If there's nothing in it for them, they simply will not improve the energy efficiency of their homes, there are just too many other more interesting things to spend one's money on, and a nice roll of fibre glass insulating quilt is very likely to miss out to the next new version of an all-singing-and-dancing mobile telephone, and with the emerging countries of the third world starting to sample those same delights, perhaps the second nail in the coffin of the Earth is already busily being banged into place.

Let us return to examining in a little more detail how we might be able to develop the use of the 'free', renewable energies that are already available to us. If the current problem in financially difficult times is getting people to pay more to improve their homes, what more can be done to assist them?

There is a very significant amount of water on the Earth, around 326 million, trillion gallons, or 1,260,000,000,000,000,000,000,000 litres. The whereabouts of it does change because it is always in a constant cycle of evaporation from the seas and oceans, raining back onto the land, and eventually flowing back into the seas and oceans. The energy which is within this water can be harnessed and used. Water is about 800 times denser than air, so even the slowest of flowing streams can yield up large amounts of energy. Harnessing this type of renewable energy is usually called hydro-electric power, and involves the large-scale construction of dams to harness river energy. The water is channelled into a turbine, which turns and creates electricity.

Energy of this type, however, need not be expensive. Micro-hydro systems, which we referred to earlier, are smaller, locally based systems which often can be used in water-rich areas to generate a remote-area power supply. A typical 5 kW small system can cost between £20,000 and £25,000, but the power is then free, and indeed since 1 April 2010 any unused power can be feed back to the National Grid for which a payment to the system owner is made in accordance with the tariff applicable at the time. Thus, after a relatively small outlay, which can be shared by several, good savings can be made.

Much the same applies to solar energy: once the cost of panels has been met, the energy is then free, and once again excess energy can be sold back for profit. The initial costs do vary, but for a standard domestic 1 kW system, supplying 825 kW hours or units of electricity into a home per annum, the net cost is currently around £5,000, saving about 468 kg of carbon emissions every year.

Also in vogue currently are heating systems that use geo-thermal energy. This involves tapping the heat of the Earth itself. The initial outlay is usually quite expensive at around £25,000 per unit for a standard size domestic home, but again, excess energy can be fed back into the National Grid, and some money can be made as a result.

So, water from the oceans, heat from the sun and heat from the ground are the Earth's offerings to try to save itself from 'us', an attempt perhaps to start banging out those nails from its coffin. It is pleasing to note that during the five-year period ending December 2009, world-wide renewable energy capacity, represented by the amount of 'kit' available to harvest them, grew by up to 60 per cent annually year on year. For wind power, capacity grew from 94 GWe in 2007 to 159 GWe by the end of 2009. Other impressive results were seen in investment in alternative sources of power: in 2007, globally, US$104 billion was invested, by 2009 this had risen to US$150 billion, a very substantial improvement. And, to a rising crescendo of applause, scientists now have a plan to power 100 per cent of the world's energy with wind, hydro-electric and solar power by 2030. Should that occur, the Earth's fight back will be complete!

The potential impact on the Earth by emerging countries is a matter of concern in that as they emerge and develop, they will want to channel their activities towards playing catch up, and that might mean a rush towards ill-informed spending on consumer goods such as computers, mobile telephones, etc. So developing renewable energies in these emerging markets will be critical over the coming years.

The economic model is of 'diminishing returns'. We might consider that in the first phase of life we are damaging the Earth collectively, then as we get older we might realise what we are doing, and in the third phase of our lives we would stop doing it and act to educate those who'd just emerged into phase one. However, there is another startling army emerging against us from leftfield, and that is that the population of the globe is growing significantly. The best estimate is that there are around 6,895,000,000 people on the Earth as you read this today. The world has experienced exponential growth in its population in recent years. Annually, around 140 million babies are born each year, and about 57 million people die during the same period.

The population is getting younger, which means that at any one time there are more people these days who are in phase one of our 'diminishing returns' model above. If this continues, those in phases two and three will eventually be over-run, and more and more damage will be caused to the Earth. These ever-increasing numbers are not sustainable, but there is very little global, co-ordinated effort at the moment to stop such growth. Despite doing its best to persuade us not to exploit its non-renewable resources, and to use its own fruits, the Earth remains in critical danger.

Worryingly, in a recent pronouncement, the Worldwatch Institute reported in its 'State of The World', 'By any measure, household expenditures, number of consumers, extraction of raw materials, consumption of goods and services, has risen steadily in industrialised nations for years, and it is growing rapidly in many developing countries.'

Despite all our efforts to the contrary, whilst we have become more and more aware of what damage we are doing (and we have done well to acknowledge that), there remain dangers that we must address. Now, this chapter is not the place for pontificating on controlling the world's population, and on making more funds available to those that can't themselves afford to build wind turbines, install micro-hydro systems, attach solar panels, or sink ground source heat pumps, but that must be the way forward and another worldwide call on those experts ready, willing and above all 'able' to contribute to the debate is well overdue.

Some good work is already underway. There are plenty of new and emerging renewable energy technologies being considered. These include cellulosic ethanol, hot-dry rock geothermal power and ocean energy. There are companies currently trying to convert biomass products into ethanol, whilst producing enzymes which contribute to a cellulosic ethanol future. This is a shift from using food crops to using waste residues and could offer a range of opportunities for farmers and others to be able to grow crops to feed the world in a sustainable way.

Ocean energy systems are about creating electrical power from the ocean waves. Presently, the potential is thought to be 'promising', but it is area specific and relies on good ocean currents. Currently, therefore, it is being concentrated in those more volatile sea areas that lie between latitudes 40 and 60 degrees. As might be expected, however, it is rather an expensive technology to work in such volatile and unpredictable areas.

Finally, one thing that we are all going to have to address is the 'not in my back yard' mentality that tends to pervade us all. It is generally accepted that a large wind turbine near one's back yard may not exactly be a thing of beauty, but just as it is surprising that a mobile phone user would complain about a phone mast being close to home, a wind turbine might just become a necessity as we all go forward into a better 'greener' future, and micro-hydro systems might become the norm just outside our pretty rural communities. After all, having arrived following a long road trip from a major conurbation, into an 'Area of Outstanding Natural Beauty' (AONB), would not the weary traveller be significantly bothered if a kettle did not yield hot water because of a lack of water supply?

If it was possible for the Earth to speak to us all, if it could be given any sort of platform to be 'heard' today, it would be likely to give a small plea and say that it 'comes in peace'; it would want to show its bandages, and it would to try to broker a deal. That deal would be to try to urge more use of the fruits that it has to offer. It would ask us all to slow down the changing pace of life, not just to give us plenty more time to look at what is going on, but also to try to reintroduce a workable harmony between the need to live and the need to make our use of the world and its environs more sustainable.

If we are not able to do that, then would the last one of us please turn off the lights when we leave. Oh! and please try not use a remote control to do so!

4 Neighbour issues

Far too often, clients and designers wait until the detailed design for the project in question is more or less finished and the process of obtaining all the necessary approvals from the relevant statutory bodies has been completed before giving any detailed consideration about the impact that the project may have upon the owners/users of any adjoining land/property. 'What's it got to do with them?' one may ask. 'Surely, once all the statutory approvals are in place what influence can the neighbours possibly have on the project?'

Of course, there are always simple projects where that might be the case; however, experience shows that a failure to take into account at an early stage in the design of the project an adjoining property owner's opinions (and even his/her practical needs) might at best be described as plain inconsiderate. Potentially, however, to fail to just stop and think for a moment about the implications of the project upon the neighbour(s), may lead to unnecessary problems further down the line.

This chapter will highlight the sort of issues – some of which may seem at first glance to be trivial – that can, given the right (or wrong, depending on one's point of view) circumstances, give rise to a potential for misunderstandings, disagreements, delays, significant levels of unnecessary costs and even the possibility of legal action. Responsibility for dealing with some of these factors may fall to the client, however it is also important that the person drawing up the works specification (whether that be the client or an appointed professional) has at least an awareness of the potential impact of these factors in order that they can, if nothing else, ensure that any necessary pre-emptive actions are taken into consideration and that they are able if necessary to properly advise their client and so discharge their professional duty of care.

These factors are discussed under five main category headings, each with its own set of sub-categories:

4.1 Property boundaries

This section on property boundaries is not intended to serve as a master-class for budding or even experienced surveyors. It is purely for the benefit of lay clients and inexperienced builders/specification writers to make them more aware of the potential hiccoughs that may beset a project if due care is not exercised in dealing with neighbour problems at an early stage. There are numerous excellent texts already in publication for the benefit of anyone wishing to gain a more detailed insight into boundary matters.

4.1.1 Title deeds

How many of us keep a reference copy of our title deeds when we buy a property? Sometimes the legal adviser sorting out the conveyance (if he/she is of a thoughtful nature) may have the presence of mind to make a copy for the client's private use, but in many cases the first time that an owner of a domestic property seeks to consult the title deeds is when the precise location of the boundary to the site is brought into question.

Of course, being nigh on impossible to replace, the original copy of the title deeds should be kept in a safe place. Believe it or not, there are still some individuals who keep their title deeds, various insurance policies, birth/marriage/death certificates for several generations, together with other important minutiae, all in a wooden box which resides in the bottom of a wooden wardrobe in the bedroom (i.e. upon a wooden first floor structure). The likely consequences in the event of a serious fire at the property are fairly predictable.

Fortunately, most title deed documents are retained by the mortgage lender in secure storage (although to reduce storage requirements, many mortgage lenders currently retain only the most recent documents, releasing the older sets to the care of the mortgagor) or, where no mortgage is held on the property, the documents may (one trusts) be in similarly secure storage, perhaps at a bank or a solicitor's office.

All title deeds are accompanied by a 'deed plan', which depicts the layout of the site boundaries – commonly by the use of a red line.

In the case of most modern properties, the deed plan will either be an extract from the developer's site layout plan (commonly at a scale of say, 1:500) or, alternatively, an extract from the Ordnance Survey map for the district at a scale of 1:1250 or sometimes 1:2500. Modern Land Registry documents (see section 4.1.6 below) use the OS mapping as the basis of their record plan.

With older properties, the deed plan can often be a hand-drawn site layout plan depicting the individual plot in relation to those surrounding it and these plans may be at larger scale, commonly as large as $^1/_{16}$ inch to 1 foot (1:192) or other metric equivalent. Occasionally, these larger deed plans may also contain some dimensional data relating to the size of the plot in addition to the common annotation confirming the approximate area of the site in square yards or acres.

The thorough reading-through of a set of title deeds has been described by some as being the most effective cure for insomnia ever invented. Of course, one could always take a professional opinion from one's legal adviser and instead enjoy falling asleep to the gentle sound of said legal adviser quietly counting his/her fees . . .

Either way, there will be occasions where it is important to establish essential information regarding the ownership and rights not only of the client, but also of adjoining property owners. In some instances, the only reliable way that this can be accomplished is by careful and thorough reference to the title deeds and deed plan (with or without professional legal advice/assistance).

Where the text of a title deed contradicts what is depicted upon the deed plan, the text of the deeds will hold sway, except in those instances where the title deeds themselves specifically state otherwise.

Another important matter relating to the site boundaries is who is responsible for the erection and maintenance of the marker on that particular boundary? Again, reference to the title deeds is the means of confirming this. Commonly, the text of the property's title deeds will state that the owner of the property is responsible for erecting and maintaining thereafter, for example, a wall of a particular material to the (let us say) southern boundary of the site and to a specific minimum/maximum height. Where the title deed text does not assist in this way, it is possible that the deed plan may provide clarification by the inclusion of an inverted letter 'T' attached to one or more of the site boundaries (see Figure 4.1a). In simple terms, the 'T' indicates that 'ownership' of a boundary feature lies with the owner within whose land it appears on the deed plan.

Of course, there are many instances where the title deed text is silent on this matter and the deed plan carries no 'T' markers. A legal opinion should ideally be sought in all cases to confirm the lay interpretation in accordance with the guidelines set out above, but especially in such instances where the deeds do not specify. It is possible that the considered legal opinion will be that the boundaries are 'party' boundaries, where each adjoining owner has a shared responsibility for the erection and maintenance of the boundary marker. Indeed, in certain cases such 'party' boundaries are specifically named in the title documents and may include 'party walls' separating adjoining properties as shown in the example at Figure 4.1(b) below.

4.1.2 Determining where the boundaries are

There are numerous matters encountered in one's day-to-day life that may appear to be very straightforward but which are found in practice to be anything but. Establishing the precise location of a property's boundaries can, in certain instances, fall squarely under that description. The deed plan may well seem easy to comprehend, after all, the red boundary lines (reproduced as thick grey lines in the monochrome illustrations below and overleaf) are drawn beside the black lines on the Ordnance Survey map are they not? How complicated can that be?

Unfortunately, problems do occur due to a number of reasons, examples of which are discussed below.

Scaling off an Ordnance Survey map to determine the actual boundary positions is to be avoided, as the mapping data (especially modern electronic mapping data, which is

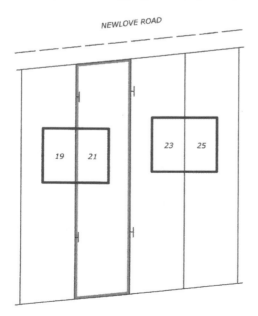

Figure 4.1(a) Example deed plan with 'T' marks

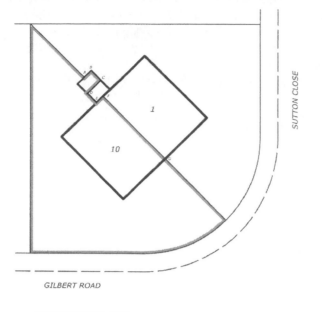

GILBERT ROAD

ABCDEFG are Party Walls

Figure 4.1(b) Example deed plan with party walls marked

sometimes based upon aerial or satellite photography) is rarely accurate enough for this purpose. Also, a black line that is 0.25 mm in width printed on a copy of 1:1250 scale map represents a line about 300 mm wide 'on the ground'.

Deed plans for modern houses on larger estate developments are commonly prepared using the developer's site layout plan, on which all of the plots are drawn out neatly and with regular sizes and ninety-degree corners, etc. Regrettably, that plan may then be interpreted on-site by an individual of less exacting demeanour, possibly using accurate instruments, but often by eye and with just a non-metal tape measure. No doubt in some odd instances guesswork plays a large part as well. Thus, if an accurate as-built layout plan is not used (assuming one such is actually prepared) the deed plan is unlikely to be a true representation of what actually exists.

There will be instances where the deed plan carries specific measurements setting out the plot length/breadth, etc. but these can sometimes be misleading. There are numerous examples from the author's own experience where the deed plan clearly stated plot dimensions that failed to bear full resemblance to the actual size of the plot as measured on site. Often, this may be because one or other property owner in the past has 'adjusted' the position of a boundary marker.

An example is the case where a surveyor accepted instructions jointly from the owners of two adjoining domestic properties, whose interpersonal relationship had regrettably deteriorated to the point that they were only communicating through their respective solicitors' offices. The instructions, prepared jointly by the two solicitors, no doubt seemed straightforward enough as both properties' deed plans were available, and in each case the deed plan clearly stated that the side boundary in question that separated the two properties was a distance of precisely 8 feet offset from each gable elevation. The first measurement taken on site was to confirm the total separating distance between the gable elevations, which was 16 feet 6 inches – back to square one! Thus, whilst having a deed plan with precise measurements annotated on it may seem to be an excellent reference, it is not always the case that this would be sacrosanct in practice.

The location of permanent boundary markers 'on the ground' is generally regarded to be the more reliable means of determining the precise position/alignment of a site's boundaries. Care must be taken, however, to determine where possible the longevity of the boundary marker in question (see section 4.1.5, Adverse Possession). An example of a potentially reliable boundary marker would be an old pre-cast concrete post – a remnant of a much older (possibly original?) post and wire fence – that is found buried within a thick, mature hedge (see Figure 4.1(c)) where the hedge has been planted to one side of the fence but over a period of time has perhaps grown more toward the adjacent property – in which case, the position of the fence posts would be highly likely to be confirmed as the true boundary line and not the centre-line of the hedge root stock.

The reason why the determination of an accurate boundary position is important is plainly to avoid the possibility of future delays, costs and even legal action for alleged trespass and/or damages. As a general rule, it is strongly advised that any potential uncertainty over the true boundaries of a site for a proposed building project should be determined prior

Figure 4.1(c) Photo of concrete post within hedge

to commencement on site and, preferably, prior to commencement of the detailed design of the project. Legion are the stories of buildings having to be thoroughly and expensively re-designed because the building would not fit into the actual plot, simply because no-one thought to confirm at an early stage whether the boundaries shown on the site layout plan and subsequently relied upon by the design team were in fact accurate.

4.1.3 What constitutes a boundary marker?

A permanent boundary marker is most commonly held to be represented by a wall or a stout fence. There are some instances where a hedge might be classified as an acceptable boundary marker but, in the author's experience, these are commonly restricted to agricultural applications and not generally to boundaries between, for example, adjacent domestic, commercial or industrial properties.

Construction joints between different forms of paving or panels of similar paving material laid at differing times by the respective property owners would not normally constitute a permanent boundary marker, irrespective of their ultimate accuracy.

4.1.4 Boundary disputes

One of the best pieces of advice that a professional can give to a prospective Client, who is about to embark upon a boundary dispute is . . . 'If at all possible, try your very best to avoid doing so'. That may seem at face value to be a rather negative approach, but it is truly a fair reflection of the experience of many professional surveyors and legal advisers.

That having been said, there are occasions from time to time when an unlucky property owner on the one part may be faced with the prospect of having to defend an unwelcome challenge from an adjoining property owner on the second part, who holds a contrary opinion as to the position and/or status of a particular site boundary. In such cases, the first party may have to decide at an early stage, whether to embark upon a series of actions that will inevitably incur him/her with substantial professional fee costs, and without any real certainty that those costs will ultimately be recoverable from the second party, whether or not their defence against the action is successful.

The alternative – capitulation to the second party's opinions – is in many cases equally unpalatable. The files of professional surveyors the length and breadth of the country are littered with cases where a client decided to fight the neighbour 'as a matter of principle'; but an experienced practitioner will tell any client that 'a point of principle' may turn out to be a very expensive luxury in the long-term. Even comparatively minor domestic boundary disputes can ultimately end up leaving even the 'successful' party hundreds, if not thousands of pounds worse off in un-recoverable legal fees.

Back to the original point – how to avoid boundary disputes. As discussed earlier in this section, if one party is determined to stick to their 'guns' it will possibly require the other party to capitulate fully to their arguments if a formal dispute is to be avoided. It is for this reason that it is important – possibly even fundamentally vital in some instances – that where a project is perceived to be likely to have an impact upon the position of a boundary or perhaps, upon the nature/construction of a boundary marker, that the client (preferably, for domestic properties) or failing that, the client's representative (contractor, designer, agent) makes early contact with the adjoining property owner to make them aware of the issues and to try where possible to resolve them amicably by whatever means are appropriate.

4.1.5 The Land Registry/adverse possession

Following a thorough review, the Land Registry's operations were the subject of new legislation contained in the Land Registration Act 2002 and the Land Registration Rules 2003. The impact of this new legislation was widespread, but in relation to boundary matters insofar as they may affect a project, there are two areas of interest to the reader of this book. These are the ways in which the Land Registry records may assist in determining who owns a particular property and what effect the new legislation has had upon 'the law of adverse possession'.

Obtaining details about the ownership of a parcel of land is a relatively simple exercise and may be accomplished online via the Land Registry website (currently www.landregistry.gov.uk and follow the links to 'find a property'). Searching for property data for domestic premises or any numbered commercial or similar premises is straightforward in the majority of cases although searches for data with regard to, for example, parcels of undeveloped land will require some knowledge of the approximate boundaries of the site. The cost of obtaining data for a single domestic property (including an extract copy from the latest conveyance together with a copy of the Land Registry location plan) currently costs just a few pounds.

The problem comes when a property is not recorded with the Land Registry. For domestic properties this is commonly because the property has not changed ownership for many years and so would not have become compulsorily registrable. In such cases, determining the ownership of a property or parcel of land becomes more complex and professional assistance will probably be necessary to achieve a positive end result.

Adverse possession is a process by which premises or property can change ownership under common law. By adverse possession, title to another owner's property can be acquired without having to necessarily compensate the other owner. This is achieved by occupying the property in a manner that conflicts with the true owner's rights over a given period of time. The Land Registry has published detailed Practice Guides in relation to this and many other topics relating to property law and the reader is recommended to study those Practice Guides to obtain a fuller understanding of the law of adverse possession and how it, in turn, relates to the Land Registration Act 2002. Schedule 6 of that Act sets out in detail how the current 'regime' is intended to operate.

It is important to recognise that the application of the Land Registration Act 2002 and adverse possession law will vary, dependant upon whether the property in question comprises registered or unregistered land.

4.2 Rights of access

It is a simple thing, but one that a surprising number of even experienced professionals manage to get wrong from time to time – put simply, is there sufficient access to/from the site to permit safe, easy access by mechanical plant etc.? Is there somewhere that a delivery wagon can safely park to unload materials and plant? Equally, is there enough room on site to safely and securely stockpile materials? And is there sufficient safe access all around the working area to carry out the construction works, perhaps to include the erection of scaffolding?

Taking the preceding set of questions to the next level – and thus, bringing matters into the context of this chapter – is there sufficient space to accomplish all of the above without having an undue effect upon neighbouring properties or their owners, thus leading to potential delays and additional unforeseen or unnecessary costs?

4.2.1 Consult the title deeds

Much as in the preceding section of this chapter, consulting the title deeds and the deed plan is a useful starting point, not only to try to establish to actual position/alignment of the site boundaries, but also to establish whether there are any specific rights of access onto adjoining land or indeed by other parties onto the project site.

Such rights of access, called 'easements', occur commonly when it is desirable to have a shared access (perhaps a path or driveway) that is intended to be used by both property owners. As an example, the boundary line perhaps may pass directly up the centre-line of a 0.90 m wide path and thus each property would own a 0.45 m wide portion of the path, but the other party would have a legal right of access to 'trespass' onto the other party's half of the path.

Another common easement example concerns older terraced properties, where a centrally placed access passageway is intended to provide rear access to the whole terrace of, say, six properties. This access would be along the passage and across the rear yard of each of the four central properties. Thus, whilst the passageway itself and the paths across the rear yards would belong, in all probability, to the individual properties on whose land they stand, the other property owners would have a right of access by virtue of an easement written into the property title deeds to pass unhindered along the passageway and across their neighbour's rear yards. Such easements are commonly displayed on the deed plan for a property using coloured shading for the respective areas of access, as depicted at Figure 4.2(a) below.

Beware the situation, therefore, where it is proposed to obstruct (either temporarily or permanently) an area where a right of access exists in favour of an adjoining property owner, for fear of suffering legal action against the owner of the site. A temporary obstruction might

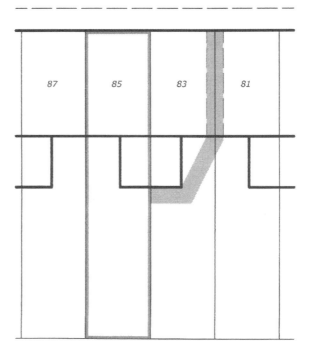

Figure 4.2(a) Example deed plan with coloured shading

be acceptable to an adjoining owner, possibly subject to a formal licence (see section 4.2.3) but it may be very difficult to enforce such restrictions to access against an adjoining owner who for whatever reason has decided not to be co-operative on the issue.

So, once again, much as in the previous section of this chapter, the essential message here is to try wherever possible to maintain the best possible working relationship with the owner of the adjoining property, or else suffer the inconvenient (and sometimes, financially punitive) consequences.

4.2.2 Access to Neighbouring Land Act 1992

The Access to Neighbouring Land Act 1992 can in some instances be a useful means for a property owner to enforce access onto land belonging to another owner. Essentially, a property owner, having formally requested permission to enter the adjoining owner's land to carry out work to this own property, and having been refused permission to do so, can apply to the court for an order enforcing the adjoining owner to permit such access.

However, it is important to remember that this Act only applies where the proposed works are for the repair or maintenance of an existing building and specifically not to facilitate carrying out any new building works. Also, a successful application for a court order may be subject to certain restrictions as to the hours of working, duration of the access, etc. In cases relating to non-domestic property the costs associated with obtaining such an order are likely to be significant.

Please note that this section should be read in conjunction with section 4.3.9 of this chapter with regard to the provisions of Section 8 of the Party Wall etc. Act 1996, which may give an alternative way of enforcing rights of access onto neighbouring land for certain works that are notifiable under that Act.

4.2.3 Erection of working platforms/scaffold licences

As discussed in the previous section, there is no guarantee of access to an adjoining owner's land – for the sake of clarification, 'access' is taken to include the erection of a scaffold platform. Assuming that the owner of the adjoining property has indicated consent to give such access, it is possible that it may be granted only subject to the parties entering into a formal agreement.

A common method of achieving such agreement, then, is to draw up a Scaffold Licence. Conventionally, both property owners would be signatories to the licence, however there are circumstances where it may be advantageous to make the contractor a signatory to the licence also.

An obvious example of why that is often preferable is to enable the owner of the adjoining property to be able to maintain a degree of control over the period of time during which the access may continue. A contractor who is a signatory to an agreement which places him under, for example, a financial penalty in the instance that the works over-run for any reason, will perhaps be more inclined to ensure that the works are completed promptly.

Other elements contained with a typical Scaffold Licence are in relation to a commitment to remediate any damage caused to the adjoining property and, in that same regard, the licence is often accompanied by a Pre-Works Schedule of Condition of the adjoining property (see sections 4.3 and 4.6).

Thus, the intention is to give the owner of the adjoining property sufficient assurance that the matter will be dealt with professionally, sympathetically and that he/she will maintain control of the proceedings throughout, should the other party fail to honour their element of the agreement.

The reader is recommended to seek appropriate professional advice before drawing up and agreeing the terms of a Scaffold Licence Agreement.

Please note that this section should be read in conjunction with section 4.3.9, with regard to the provisions of the Party Wall etc. Act 1996, which may give rights of access onto neighbouring land for certain works that are notifiable under that Act and thus obviate the need for a Scaffold Licence.

For works requiring the erection of a scaffold or similar platform that will obstruct the highway, the reader should refer to Chapter 1 of this book.

4.3 The Party Wall etc. Act 1996

This legislation is considered by many to be the single most significant factor affecting minor construction works projects in England and Wales (outside of the Central London Boroughs) that has been introduced over the past 25 years. It is significant simply because of the potential financial and also time-delay implications that can arise from its application.

Again, as with preceding sections of this chapter, it is not the intention here to provide a detailed working knowledge of the Party Wall etc. Act 1996 – there are numerous subject-specific works already in existence. The intention is to try to provide a less-experienced person (be that a client or a professional) with an awareness of the basic facts and, hopefully, some assistance in identifying and avoiding potential pitfalls arising from the application of the Act to any given project. As always, there can be no substitute for availing oneself of sound advice from an experienced practitioner in this specialist field at an early stage in any project.

4.3.1 Brief history of the Act

Laws concerning party wall matters were in existence before the Norman Conquest of 1066 and one of the established professional forums for party wall matters makes reference in its own name to the mythological story of the Babylonian lovers Pyramus and Thisbe, who communicated with each other through a crack in the party wall separating the properties of their two families – who were not at all content with the burgeoning relationship of these two ill-fated lovers.

The first parliamentary legislation controlling party wall matters was the Charles II Act of 1666, which was 'An Act for rebuilding the City of London'. However, the current legislation contained with the Party Wall etc. Act 1996 (which for ease of reference we shall refer to for the remainder of this section simply as 'the Act') is generally held to be based upon Part VI of the London Building Acts (Amendment) Act 1939, entitled 'Rights etc. of Building and adjoining owners'. These London Building Acts only pertained to the inner London Boroughs and so the rest of England and Wales had to wait for detailed legislation on the issue of party walls until the inception of the Act on 1 July 1997.

4.3.2 Terminology used in the Act

In earlier sections of this chapter, repeated reference has been made to the 'legal boundary' separating adjoining properties. However, the Act refers instead to the 'line of junction' and it is important to differentiate between these two terms. In the context of the Act, the 'line of junction' may most commonly represent, say, a horizontal line on the ground that does not necessarily also project in a vertical plane, as would be the case with the 'legal boundary'.

Less commonly, however, the Act may also relate to works to a 'party structure', that could be an intermediate floor separating, for example, two apartments in which event the line of junction would be located somewhere within the plane of the separating floor structure.

Two other important terms used in the Act are the nomenclature of the parties involved. Specifically, the property owner who desires to carry out building works subject to the provisions of the Act is always referred to as the 'building owner'. The owner of the neighbouring property potentially affected by the proposed works is always referred to by the Act as the 'adjoining owner'. There may, of course, be more than one individual person or entity for each of the above terms where properties are in multiple ownership/occupancy.

A 'party wall' is defined by the Act as:

(a) a wall which forms part of a building and stands on lands of different owners to a greater extent than the projection of any artificially formed support on which the wall rests; and

(b) so much of a wall not being a wall referred to in (a) above as separates building belonging to different owners.

Typical examples of party walls are shown in Figures 4.3(a) and 4.3(b).

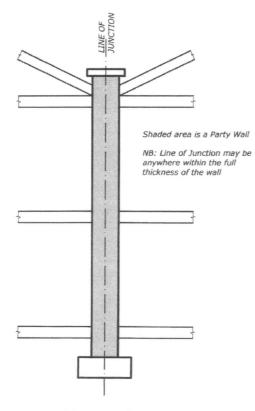

Figure 4.3(a) Party wall type A

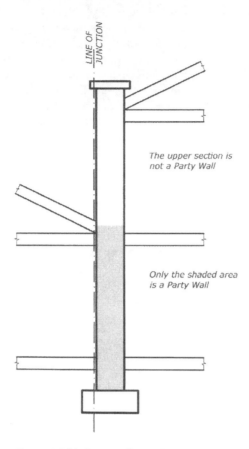

LINE OF JUNCTION

The upper section is
not a Party Wall

Only the shaded area
is a Party Wall

Figure 4.3(b)   Party wall type B

In Figure 4.3(b), although the external wall of the two-storey building on the right stands entirely within the line of junction, the shaded portion of the wall is nevertheless a party wall over that portion of its area where the building to the left subtends it. Similarly, if the properties were 'staggered' or offset horizontally front to rear, then only that section where the two buildings abut would be a party wall.

The Act also refers to a 'party fence wall' (see Figure 4.3(c)) which is taken to mean a wall that does not form part of a building, but which stands on lands of different owners and is used for separating the adjoining lands. This would of course include a garden wall where it stands astride the line of junction.

Importantly, it should be noted that the term 'party fence wall' does *not* include a boundary wall constructed wholly on the land of one owner, even if its 'artificially formed support' (i.e. its foundation) projects beyond the line of junction into the land of another owner (see Figure 4.3(d)).

Where a dispute arises under the Act (see section 4.3.7 below) the parties either appoint one 'agreed surveyor' or two 'surveyors' to administer and resolve the dispute. In this regard, the terms relate to an individual who is, ideally, experienced in the application of the Act but who in any event cannot be one of the parties to the dispute. The individual appointed

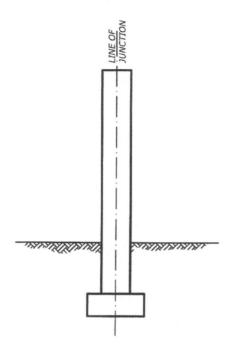

Figure 4.3(c) Party fence wall

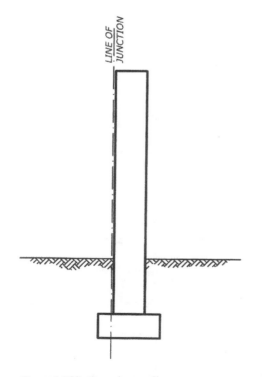

Figure 4.3(d) Boundary wall

could indeed be a practising surveyor, but would be just as likely to be a building engineer, structural engineer, architect or other building professional who has suitable experience in the application of the Act.

There are three principal elements to the Act that may affect a given project and these elements are discussed in the following three sub-sections of this chapter.

4.3.3 Works to erect new structures on the line of junction

Section 1 of the Act applies where a building owner intends to build on any part of the line of junction. However, section 1 only applies in the instance that either:

(a) the lands of the different owners are not already built on at the line of junction; or,
(b) they are built on at the line of junction only to the extent of a boundary wall (not being a party fence wall or the external wall of a building).

In other words, if the building owner intends to erect a new party wall or party fence wall, or indeed, if the building owner wants to build, for example, a new structure (e.g. an extension to a dwelling), with the outer face of the side elevation wall built on the line of junction but wholly within his own land, then those works will fall subject to Section 1 of the Act.

As an example of how this rule may be interpreted, Figure 4.3(e) depicts a property with an existing extension to the left-hand side, where the upper storey is supported by four pillars of masonry, which form a semi-open 'carport' to the ground storey. The owner of the property now wishes to infill between the pillars to form an enclosed 'garage', as shown by the shaded sections of walling. Where the pillars stand, the land is 'already built on at the line of junction', however the proposed works to erect the infill sections in between the pillars would require prior service of a notice under Section 1 of the Act, because over those three sections the lands are not built on at the line of junction. Depending upon the intended depth of the proposed infill wall's foundation excavation relative to the foundation of any adjoining structure, a notice under Section 6 of the Act (see section 4.3.5 below) may still be required.

Of course, if the wall at Figure 4.3(e) were to be erected only a short distance inside the line of junction, then Section 1 of the Act would not apply. One may ask of course, 'precisely how far away from the line of Junction would the wall have to be, for Section 1 not to apply?'

In response to that query, it is perhaps important to note that the Act does not refer to works 'near to' or 'adjacent to' the line of junction. Rather, it only uses the word 'on' the line of junction and so perhaps the only argument that may be applied here relates to the actual accuracy with which one can safely determine the 'precise' location of the line of junction. Thus, if one could be certain of the actual location of the line of junction to an accuracy of, say, one millimetre, then it follows that works more than one millimetre distant from the line of junction would fall outside the scope of Section 1 of the Act. Unfortunately, it is rarely the case that one may be this accurate in establishing the position of the line of junction and so pragmatism must surely rule the day.

This issue has long been the subject of lively discussion between party wall professionals and the reader is best advised to take specialist professional advice on the issue before embarking upon any works in close proximity to a site boundary.

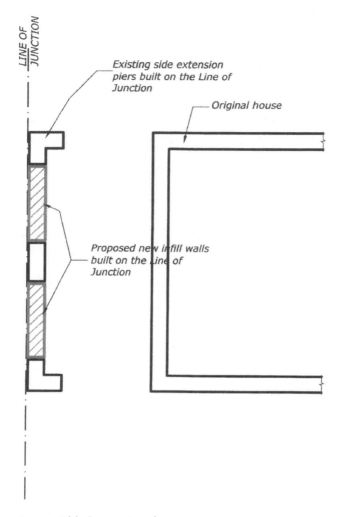

Figure 4.3(e) Section 1 works

4.3.4 Works affecting existing party structures

Section 2 of the Act applies where at the line of junction, a party wall exists. Section 2 also applies where at the line of junction, a boundary wall exists, so long as that boundary wall is either a party fence wall or, alternatively, is the external wall of an adjoining owner's building.

The Act sets out in detail at Section 2(2)(a)–(n) the various examples of the different ways in which this section of the Act may apply. It is not the intention to repeat the precise descriptions within the Act, as these may be easily obtained and studied in detail by the reader at his/her leisure, however a non-comprehensive summary of the most common examples that occur are as follows:

- Underpinning or raising a party wall.
- Carrying out repairs to a party wall.

- Demolishing and rebuilding a party wall that is not fit for purpose.
- Cutting into a party wall for any reason (which may also include for the insertion of a damp-proof course).
- Cutting away of any projecting footing, chimney breast, flue or jamb (subject to making good).
- Cutting away any overhanging part of the adjoining owner's property (subject to making good and providing adequate weathering).
- Exposing a Party wall or party structure that was hitherto enclosed (subject to making good and providing adequate weathering).

4.3.5 Excavation works

Section 6 of the Act relates to works of excavation – which may be for any reason – within influencing distance of a neighbouring building or structure. The phrase 'within influencing distance' is not drawn from the Act but is the author's own and has been used in this context because it reflects the intent and purpose of this section of the Act, which is to seek to protect the adjoining owner's property from the unnecessary risk of damage due to instability as a result of unlawful or ill-conceived close-proximity excavations by neighbouring property owners.

It has been noted by many professionals since the inception of the Act in 1997 that many property owners (indeed, also some builders and other professionals) still do not appreciate that works of excavation may fall subject to the Act, even when those excavation works may not directly affect a party wall for example.

Section 6(1) of the Act states two circumstances where the provisions of the Act will apply in regard to excavations, these being:

(a) Where a building owner proposes to excavate within a distance of three metres measured horizontally from any part of a building or structure belonging to an adjoining owner; and where any part of the proposed excavation will within those three metres extend to a depth lower than the underside of the foundations of the adjoining owner's building or structure.

(b) Where a building owner proposes to excavate within a distance of six metres measured horizontally from any part of a building or structure belonging to an adjoining owner; and where any part of the proposed excavation will within those six metres meet or extend below a plane drawn downwards at an angle of forty-five degrees from the line formed by the intersection of the plane of the level of the bottom of the foundations of the building or structure of the adjoining owner with the plane of the external face of the external wall of the building or structure.

There are numerous examples of diagrams depicting the relationship of excavation with the adjoining owner's structure; however, the diagram at Figure 4.3(f) serves very well to simplify the matter. In essence, any excavations that will cut through into the hatched area below the thick line will require service of Notice under section 6 of the Act.

As inferred by the opening paragraph of this sub-section, the purpose of this part of the Act is to prevent a building owner from excavating, say, for a new 3 metre deep basement in close proximity to an adjoining owner's building or structure, without demonstrating to the adjoining owner's satisfaction that adequate precautions have been taken to prevent instability and

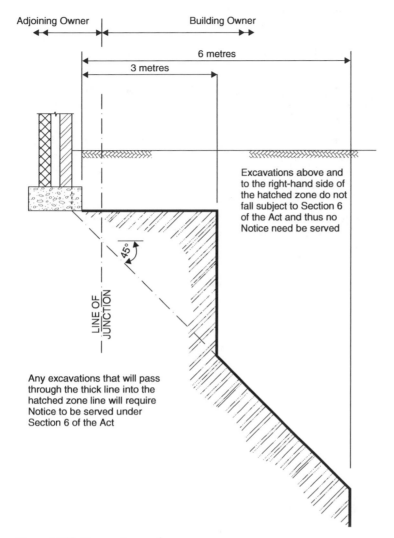

Adjoining Owner | Building Owner

6 metres

3 metres

Excavations above and to the right-hand side of the hatched zone do not fall subject to Section 6 of the Act and thus no Notice need be served

45°

LINE OF JUNCTION

Any excavations that will pass through the thick line into the hatched zone line will require Notice to be served under Section 6 of the Act

Figure 4.3(f) Excavation works

movement occurring to the adjoining owner's property. Inevitably, there are likely to be any number of instances where excavations are proposed to take place at, say, 2.9 m distance away from an adjoining owner's structure and to a depth of perhaps only 100 mm lower than the foundations of the structure. In the vast majority of cases this would represent a minute level of risk of instability or movement and yet, Section 6 of the Act would still apply.

Excavations within six metres horizontal distance will only generally fall subject to Section 6 of the Act in cases where very deep excavations are necessary or more commonly where, for example, piled foundations are proposed. There are differing views amongst experts in the application of the Act as to whether piling techniques using 'driven piles' (as opposed to bored or augured piles) truly constitutes works of 'excavation'. However, there is an argument that even driven piles will result in some displacement of the ground and that this must therefore be a process of excavation and, thus, subject to Section 6 of the Act. Works of excavation might also therefore include, for example, site investigation works involving the sinking of boreholes.

4.3.6 Notices

The Act demands that, in the event that the proposed works fall subject to the provisions of Section 1, 2 or 6 of the Act, then formal written notice must be served upon the adjoining owner(s) by the building owner. For works subject to Section 1 or Section 6 of the Act, that notice must be served not less than one full calendar month prior to the intended start date of the element(s) of the proposed works that fall subject to the Act. For works subject to Section 2 of the Act, the required notice period is two calendar months.

The purpose of serving a notice is not only to formally notify the adjoining owner of the proposed works, but also to take advantage of the rights given by the Act. It is very important to recognise that, until such time as the building owner serves a notice on the adjoining owner(s), the Act cannot be invoked, thus preventing either owner from taking advantage of the rights given by the Act.

Such rights may include, for example, a right to enforce access onto the adjoining owner's land in order to carry out some of the works (see section 4.3.9 below) in cases where the Access to Neighbouring Land Act 1992 is of no assistance (see section 5.2.2 above).

There are numerous 'standard forms' of notice available either from government websites or from other published works by or on behalf of several professional bodies. However, there is no legal requirement for such forms to be used, as the Act only requires that a notice must state:

- The building owner's name and address
- The nature and particulars of the proposed works
- The intended start date.

If the works are subject to Section 6 of the Act, then the notice must also be accompanied by plans and sections showing:

- The site and depth of any excavation the building owner proposes to make; and
- If he proposes to erect a building or structure, its site.

Clearly, however, the more detailed information that is provided by the building owner, the more likelihood there is of the adjoining owner understanding the proposals and being satisfied that suitable precautions are being taken to prevent instability and possible damage being caused to his/her property as a result of the proposed works.

It should be noted that the notice relates only to those elements of the proposed building works that are notifiable under the Act. For example, works of excavation beyond the relevant horizontal separation distance (3 metres or 6 metres as the case may be) or above the level of the adjoining owner's own building foundation may be commenced at the building owner's discretion, as they fall outside the scope of the Act.

4.3.7 Disputes

The Act states that the adjoining owner, upon receiving a formal notice from the building owner, should respond to that notice in writing within 14 calendar days stating whether he/she agrees or disagrees to the works proposed in the notice.

If the adjoining owner writes in agreement to the proposed works, then at the expiry of the remainder of the relevant notice period (i.e. one month or two months) the building owner may proceed with the proposed works – assuming of course, that appropriate approvals have already been obtained under the Town and Country Planning Act, the Building Regulations and any other relevant legislation. If a building owner seeks to commence the notifiable works before the expiry of the remainder of the notice period, this must also be by written agreement between the parties.

However, if the adjoining owner does not confirm his agreement to the works outlined in the notice within the fourteen-day period, then he/she is deemed to have dissented from the notice and a formal dispute is deemed to have arisen between the parties. It is a common misconception amongst lay persons that a failure to reply to a notice signifies consent, but this is specifically not the case in regard to the Act.

If the adjoining owner does formally confirm in writing that he/she disagrees to the proposed works, the Act does not require that any reason(s) be stated for that disagreement. Observations by practising professionals specialising in party wall matters tend to confirm that a majority of party wall disputes for domestic premises arise not due to any genuine concerns over possible structural instability, but due to more fundamental issues relating to the pre-existent interpersonal relationship between the two parties. The author is aware of at least one adjoining owner – an avid supporter of his local football club – who openly admitted that he had dissented to a notice purely because his next-door neighbour was known to support his club's main local rivals.

It is also not unknown in some cases for the adjoining owner to erroneously view the party wall notice procedure as another means of preventing the building owner from erecting their proposed structure when attempts to frustrate the building owner by means of the planning legislation have previously failed. However, it must be understood that the Party Wall etc. Act 1996 is an enabling Act, not a preventive Act.

In cases where a dispute does arise, Section 10(1) of the Act sets out the procedure that must be followed, which is that either:

(a) both parties shall concur in the appointment of a single 'agreed surveyor'; or
(b) each party shall appoint a Surveyor.

In the case of option (b) above, the two surveyors act as a joint tribunal. It is important to note that they are *appointed* by the respective building owners and not *instructed* by them. The two surveyors must act independently of their appointing owner(s), whilst at the same time seeking to protect their appointing owner's interests as defined by the Act. The two surveyors are also required, before undertaking any other task, to agree upon the selection of a 'third surveyor', usually a senior party wall expert, who will not become active in the dispute and thus will incur no fee costs unless and until such a point that the two surveyors are unable to agree upon any matter relating to the dispute. At that point, the third surveyor would be contracted by one or other party and would be called upon to adjudicate only upon any area(s) of disagreement. The appointment of the third surveyor is a comparatively rare event in party wall disputes.

The agreed surveyor or the two surveyors, as the case may be, must be appointed by the respective owner(s) in writing. Their appointment is a statutory one and their instructions, once served upon them, cannot be rescinded by either party.

Section 10 of the Act further states that the agreed surveyor or the two surveyors, as the case may be, are empowered by the Act to settle by award any matter

(a) which is connected with any work to which the Act relates; and
(b) which is in dispute between the building owner and the adjoining owner.

The award is a formal document that is prepared by the surveyor(s) and in it, the surveyor(s) may determine

(i)   the right to execute any work
(ii)   the time and manner of executing any work; and
(iii) any other matter arising out of or incidental to the dispute including the costs of making the award (see also section 4.3.8 below).

A 'typical' party wall award document will probably also have appended to it a 'Pre-Works Schedule of Condition of the adjoining owner's property (see section 4.6 below) plus the agreed project drawings and possibly also, a 'method statement' setting out the details of the procedure by which the contractor will carry out any sensitive works.

Once the award has been drawn up and signed by the agreed surveyor or the two surveyors as appropriate, it must be served upon the parties forthwith, as there is a time limit of fourteen days from the date of the award within which either party to the dispute may appeal against the content of the award in the county court, if they should see fit.

4.3.8 Costs/expenses

In the event of a dispute arising under the Act, then Section 10 of the Act states that the reasonable costs incurred in

(a) making or obtaining the award
(b) reasonable inspections of work to which the award relates; and
(c) any other matter arising out of the dispute.

shall be paid by such of the parties as the surveyor(s) making the award shall determine.

Whilst this appears at face value to imply that the surveyor(s) may apply the legal principle that 'costs follow the cause', this is rarely the case, as the entire basis of the Act is that the costs would be met by the party for whom the proposed works will provide benefit. Thus, in cases where a building owner is serving notice in relation to works to erect, say, an extension for the use of his family, it is entirely likely that the award would determine that the building owner should pay 100 per cent of the costs of making the award.

However, if the dispute was in relation to works on say, a dilapidated party wall or party fence wall, it is possible that the award may determine a split liability for the costs not only of making the award, but also for any necessary building works.

Clearly, therefore, any building owner considering undertaking building works that may fall subject to the Act is under a significant financial incentive to take care to ensure that at the earliest possible stage in the project the adjoining owner is provided with detailed, easily understood information about the proposed works, in order to reduce insofar as is possible the likelihood of a formal dispute arising.

4.3.9 Rights of access

Brief details of other relevant legislation enabling access to neighbouring land have been discussed at section 4.2 of this chapter. However, the Party Wall etc. Act also gives certain powers of access onto the adjoining owner's land. It is always preferable, of course, to be able to mutually reach a cordial agreement with an adjoining owner to have access onto his/her land, but for cases where this has not been possible, Section 8 of the Act does provide assistance in this respect – at least in regard to those elements of the works that would be notifiable under Sections 1, 2 and/or 6 of the Act.

Section 8 of the Act states that a building owner, his servants, agents and workmen may during usual working hours enter and remain on any land or premises for the purpose of executing any work in pursuance of the Act and, furthermore, it states that he/they may remove any furniture or fittings or take any other action necessary for that purpose. It is commonly taken by most professional party wall surveyors that the phrase 'and remain on any land' implies the erection of an access scaffold or other similar semi-permanent access platform. This right of access is exercisable only following service of prior notice, which in the majority of cases must be served not less than fourteen days ending with the day of the proposed entry.

This right of access has become much more important recently, as revisions to relevant Health and Safety legislation now more effectively prohibit the practice of, for example, bricklaying 'over-hand' (i.e. from inside the subject property) once the construction rises more than one storey high. It is apparent, therefore, that any project on or close to a property boundary (i.e. where there would be insufficient space to erect a scaffold between the wall and the boundary) of more than a single-storey height will be effectively unable to proceed without either the neighbour's cooperation or the imposition of rights of access under Section 8 of the Act.

It is important to note that these rights of access only apply, however, to those elements of the work that would be notifiable under the Act. Thus, if the side elevation of an extension were intended to be located other than on or astride the line of junction (even if only a very short distance away from the line of junction) then Section 1 of the Act would not apply, no relevant notice could be served and, hence, no rights of access under Section 8 would be available to the building owner.

Similarly, even if Section 6 of the Act applied and the building owner duly served notice for proposed excavation works close to a site boundary, any rights of access would apply only until such time as the excavation works were completed (i.e. building work had progressed up to ground level and the resulting trench backfilled) and not for any further works that followed.

4.3.10 What happens if no notice is served?

In an instance where works that would fall subject to the Act were commenced without prior service of a relevant notice under the Act, the only real course of action open to the adjoining owner would be to apply to the court for a restraining injunction to prevent the works proceeding until such time as a notice had been served and the resultant due process followed to a conclusion.

Such proceedings are not to be entered into lightly as the potential costs of such an action would be likely to be considerable. Furthermore, recent experience in the courts suggests that many judges would only be likely to entertain granting such an application for injunction if it could be demonstrated that the proposed works carried a high perceived risk of instability and resulting damage. Thus, in many instances, if a building owner elects not to serve a notice for works subject to the Act, there is often little that the adjoining owner can realistically do about this. However, without serving a notice, the building owner cannot, of course, then take any advantage of the rights of access granted by Section 8 of the Act.

Also, there have been a small number of cases reaching court in the more recent past of an adjoining owner suing the building owner for damage caused to his/her property in an instance where no prior notice had been served under the Act – this despite the fact that such notice was clearly necessary. The reason why this is important results from the judge's comments in at least two recent cases.

It is a basic principle of UK common law that if one person takes legal action against another, then he must prove the guilt of the person he is accusing to the satisfaction of the court. The standard of proof required is dependant upon the severity of the accusation and for the majority of civil law would be on the basis of a 'balance of reasonable probabilities'. In an instance, therefore, where an adjoining owner were accusing a building owner of having caused structural movement and damage to his/her property, it would be necessary to prove that the building owner's works were the most likely sole cause of the movement and damage – this is in many instances much more difficult to achieve than may at first glance seem to be the case.

However, experience suggests that in recent cases taken to court, where a building owner has failed to serve a notice, some judges have been inclined in certain instances to award a 'reverse burden of proof' against the building owner – thus the building owner would have to prove that the damage could *not* have been caused as a result of his/her works. This is highly likely to be equally as difficult to prove in practice.

4.4 Fire

Chapter 1 of this book deals in detail with the subject of the prevention of the spread of fire, etc., and so it is not proposed to repeat that discussion in detail here. Rather, it is intended in this section to look purely at the precautions that the client and/or project designer should consider taking with regard to adjacent properties and how this may affect the drawing-up of a specification for the works.

4.4.1 Fire boundaries/prevention of fire spread

Inexperienced designers commonly underestimate the necessary separating distance between adjacent properties to fully negate the potential for spread of fire from one property to the next. The closer to one's boundary a new building is proposed to be sited, the more expense will be incurred at the construction stage in the installation of fire-resistant materials.

Whilst most modern buildings on neighbouring sites will have been designed in accordance with the relevant contemporary Codes of Practice and Regulations, there are large numbers of older structures still in continuous (and quite legal) use that were not necessarily designed with such considerations in mind.

Thus, aside from the factors suggested at sections 4.1 and 4.2 of this chapter, the client and designer should also take note of, and make due allowance for, the proximity and apparent age of adjoining property when determining the site layout. It is possible that the relocation of a proposed building even a few metres further away from an adjoining property may enable the specification to be re-written to permit significant cost savings on fire retardant cladding materials, etc.

4.4.2 *Means of escape where this affects neighbouring properties*

Similarly, consideration must be given at the early planning stages of a project to not only the means of escape from the finished building, but also to the need to escape in the event of fire during the construction phase.

An example was a recent project to split existing industrial premises into three individual industrial units for sale and/or lease. The premises was single storey for the most part, but with a two-storey warehouse section at the rear. The premises were built tight up against the premises to either side; indeed, the property formed part of a continuous structure with the one to the right. Most significantly, the rear means of escape was shared with the premises to the right-hand side, which also belonged to the client but which, following the completion of the conversion works, would subsequently be in separate ownership.

The Building Regulations thus required a secure, independent means of escape for all three of the proposed industrial units. That would not have been a difficult matter in itself – purely a logistical exercise in creating separate compartmentation for the safe egress of the workforce out of the building. However, the complicating factor was that the rear fire escape route from the site was onto land in separate ownership, and thence through a second parcel of land in yet different ownership, before the 'refugees' from any fire would be able to access the public highway. The most difficult factor of all, however, was not the escape route, but the fact that the client had been labouring under the misapprehension that his existing easement granting legal right of egress through these other properties would be easily transferable to the new owners of the proposed industrial units. Regrettably, this proved not to be the case, and whilst the first property owner had no objections to agreeing a variation to the easement, the third property owner would not do so, as he claimed he had other plans for that portion of the site.

In the end, the third owner did agree to a variation of the easement, albeit with certain qualifying factors including a requirement for a significant financial settlement.

Unfortunately, the news regarding the limitations on the easement only came to light at a very late stage in the project (construction works were already well underway). Had the information been available at the outset, the project would have been designed very differently and ultimately, much more economically for the client. The 'moral' of the story is, 'never assume the neighbours will be happy to accommodate you'.

4.5 Environmental effects

Environmental matters are covered in greater detail elsewhere in this book and so it not proposed that this section will pass into any significant detail on the subject, except insofar as these matters might affect the relationship between the client/designer/contractor and neighbouring landowners.

4.5.1 *Noise pollution*

There are some of us in the construction industry who still remember a time when the only musical sounds escaping from a construction site emanated from the lips of the itinerant plumber/joiner/bricklayer/plasterer who coincidentally considered himself to be a passing balladeer (NB: the reference to only the male gender is considered acceptable in this context, for this was surely a time when the presence of female construction operatives was virtually unknown on a building site and so the requirements of political correctness are still satisfied). So long as said balladeer was reasonably tuneful and excepting that he avoided use of 'alternative' lyrics, the practice was rarely, if ever, a cause of significant noise pollution.

Oh, that it were still the case. At the risk of being consigned to the ranks of 'grumpy old men' why, one may ask, is it necessary in the twenty-first century for building contractors' operatives to need to be entertained throughout their working day by the incessant blaring of contemporary music, at volume levels more normally encountered within the confines of a licensed premises or nightclub? It was even argued by one site worker that he had turned up the volume on his 'ghetto blaster' because he couldn't hear the music through his ear defenders!

Thankfully, many of the more enlightened medium- to large-sized contractors now prohibit the playing of loud music on their sites. Aside from Health and Safety concerns related to the practice, these companies are recognising at long last that the noise pollution effects upon neighbouring property owners can ultimately lead to resultant and unexpected delays and costs.

However, many small- to medium-sized building contractors still pursue the practice. There are obvious situations where it would be unthinkable to have loud music blaring – for instance in close proximity to a hospital or school. But many fail to have any consideration for the poor individual living next door to a domestic project building site, whose only recourse to prevent the noise is either contacting the local authority Environmental Health Officer or, alternatively, to consider legal action. The latter will certainly be very expensive and not guaranteed to be successful and either method will be likely to put undue pressure upon the parties' ability to enjoy a mutually beneficial interpersonal relationship as neighbours.

So, when writing a works specification, it may be appropriate in more cases than is immediately obvious to include a clause prohibiting the playing of loud music during working hours.

Of course, building sites do produce significant levels of noise pollution anyway and this is generally unavoidable to some degree. There is legislation to limit the noise levels generated however and most local authorities currently display on their respective websites details of acceptable hours during which it is expected that noise may be generated. Again, there is no reason why reference to these limits should not be included in any works specification.

4.5.2 Airborne pollution, water-borne pollution

Again, compared with as little as twenty years ago, the legal requirements for control of material pollution within and emanating from building sites have become much more rigorous. The details of the relevant legislation are discussed elsewhere in this book.

It should be remembered, however, that irrespective of any specific laws governing building sites, the discharge of, say, surface water from one parcel of land onto an adjoining parcel of land would constitute as act of nuisance, under common law. The same would no doubt also apply to the discharge of airborne dust or water-borne pollutants onto neighbouring land.

Thus, the building works specification will need to reflect these possible issues. Aside from specifying protective screens and the like when airborne dust is likely to impact on a neighbour's property, discuss the issues first with the neighbour and then include in the specification a requirement for the contractor to carry out a reasonable sweep-down at the end of every shift and also budget for the likely cost of paying for the neighbour's windows, etc., to be cleaned, say, once a fortnight or even more regularly than this if the situation requires it.

It must be remembered, however, that even if the contract documents and specification place an onus upon the contractor to keep both his site and neighbouring properties clean, it will still remain the responsibility of the property owner to ensure that this is adhered to.

Any adjoining property owner seeking redress for such pollution will address any such claims directly to the property owner, who will be potentially liable for any resultant damages.

4.6   Pre-Works Condition

Reference has been made several times during this chapter to 'Pre-Works Condition Schedules'. It is becoming more commonplace for more enlightened building contractors to either produce such a schedule themselves or, alternatively, to engage to services of an independent third-party professional to do this for them.

4.6.1 *The inspection and schedule*

'Carry out an inspection of the adjoining owner's property and produce a schedule of the condition of the property' On the face of it, this sounds like a fairly simple task – and so it is, so long as certain simple guidelines are carefully followed. These are:

- Take your time
- Include all potentially affected areas
- Be accurate
- Be concise
- Be complete
- Cross-reference text to photographs
- Include a location plan on larger properties.

Many building professionals allow insufficient time to properly undertake a condition survey of a property. Make sure that you have time to examine each surface (wall, floor, ceiling, etc.) of every room and external elevation, not forgetting external paved areas, etc.

Many professionals will tend to include in the Condition Schedule only those parts of the property that they believe are potentially likely to be affected by the proposed works. However, more experienced professionals will know that occasionally, particularly when dealing with excavations in poor ground conditions, damage can sometimes be caused even on the opposite side of an adjacent building, furthest from the open excavation.

There is little point in undertaking a Condition Inspection unless the data recorded in the ensuing schedule is accurate, concise and complete. For example, avoid simply referring to the presence of, say, 'a fracture to the wall/ceiling plaster'. Instead, describe the start position of each individual fracture, plus its orientation, direction, length, finish position (if appropriate) and, where possible, its approximate width.

Fracture widths can for ease of reference simply be classified in accordance with the Building Research Establishment (BRE) Digest 251:1995, entitled 'Assessment of damage in low-rise buildings'. Table 1 in BRE 251 sets out six classification 'grades' for visible damage to walls, with particular reference to the ease of repair of plaster and brickwork or masonry. It is quite acceptable within a Condition Schedule to classify the fractures/cracks either by the numerical grading system from '0' to '5', but it is often easier for lay clients to understand the use of descriptive terms from BRE 251 such as 'hairline cracks', 'fine cracks', etc. These two terms will be found to cover the majority of visible cracks in most domestic environments, which will rarely exceed 1 mm in width. Note that any fractures greater than 1 mm in width should be brought to the attention of the client before work proceeds, as they may be an indication that some degree of instability already exists at the property being inspected and it is therefore possible that some additional preventative measures may need to be considered.

Condition Schedules should comprise not only a detailed written narrative describing the defects present, but also a comprehensive photographic record. Of course, some minor fractures will not be visible on all but the most high-resolution photographs. Each photograph should be cross-referenced to the narrative. In these environmentally sensitive times, it is rarely acceptable to print off countless pages of large-size reproductions of the photographs and so a more responsible course of action is to print small-size copies (perhaps 9 or 12 to a page) but also include electronic copies of the photographs on a CD-Rom or other electronic media.

For most domestic properties, the above level of detail will be quite adequate. However, for larger premises it is usually better to annotate each individual defect and to additionally record the location of each defect on a layout plan of the premises. Using this method, it was possible some years ago for the author to record the precise location of more than three thousand individual defects spread liberally around a large hospital complex comprising more than twenty separate buildings that were the subject of a claim for mining subsidence. The level of detail recorded subsequently was a primary factor in the claimant agreeing to reduce the original claim of £7.2million down to a settlement of just £325,000.

4.6.2 The benefits

The Condition Schedule can, in certain instances, provide the best possible means of two adjoining property owners maintaining a good interpersonal relationship in the event that damage may be caused during the building owner's construction works.

Most professional engineers and surveyors will testify (not as qualified psychologists, but perhaps as amateur observers of human nature) that the majority of property owners (especially domestic property owners) do not as a rule make regular inspections of their property to see if any cracks may have developed. That is, they do not do this unless some outside 'catalyst' comes into play, such as a neighbour excavating a trench for his extension's foundations alongside. Once inspired to inspect their property by such a 'catalyst', the discovery of any sort of damage results in the individual reaching for the 'fine toothed comb', whereupon they proceed to find yet more and more evidence of 'new' fractures.

In truth, it is more likely than not that the fractures will have been present for some considerable time and are quite unrelated to whatever 'catalyst' prompted them to inspect their property in the first place.

Notwithstanding the likely truth of the preceding paragraph, the individual concerned will commonly resist all attempts to shake them from what they have decided to believe to be 'true'. The likely consequence, if there has been excavation work to the neighbouring property, will be an argument over the true cause of the damage and, irrespective of the final outcome, two neighbours who might otherwise have enjoyed many years of peaceful coexistence, are now at 'DEFCON 3'.

This whole situation could probably have been easily avoided, if a Pre-Works Condition Schedule had been prepared, as this should have identified all pre-existent defects, thus leaving the two parties to then simply agree that any new defects not recorded by the Schedule immediately prior to commencement of the excavation works are therefore likely to be due to that cause (assuming that some other significant 'catalyst' has not coincidentally occurred in the meantime – such as the UK earthquake that occurred in the early hours of 27 February 2008). All that would remain for the parties is to agree the means and cost of repairing the 'new' damage.

5   Minor works

Foundations

All structures require an adequate foundation, which is needed to transfer the weight of the building onto the earth below. There are various types of foundations that can be constructed, and their depth is dependent on:

- The ground conditions, including type of soils, difference in ground levels.
- Location, type and height of trees.
- Depth of existing or proposed drainage.

Where foundations are formed in poor soil, are close to existing trees or are within the vicinity of deep drains, then an engineer may be required to design the foundations.

Foundations need to extend beneath the 'frost line', which is normally below 750–1,000 mm in depth. They should also be formed in suitable soils, such as clay, ballast, chalk or rock. Poor soil could be earth mixed with rubble or imported soil, which is considered 'fill ground'. Sand could also be considered poor soil as well as ground near mine shafts or natural openings within the soil such as 'shallow holes' within chalk areas.

Foundations within highly shrinkable soils, like clay, should be taken to a depth of 600 mm below live tree root activity. Stripping the husk from the root can identify an active tree root and if it reveals a moist white fibrous piece, then the root is active in removing the water/moisture from the clay. Once the moisture from the clay has been removed, the clay loses its characteristic properties of being able to be mould into ball and become what is called 'desiccated'. This desiccated clay is unable to support any load placed onto it and as a result it creates settlement.

Different types of trees have different effects on foundations. British Standard 5837 suggests that special precautions should be taken (unless investigation shows them to be unnecessary). It offers a user-friendly guide to assess the depth of foundations to different types of mature trees against the distances from the foundations. It should be noted that the removal of a tree that might be considered in the way would result in the condition called 'clay heave'. This is where the clay bulks as a result of the swift removal of a tree and where the balance between tree and soils has been interrupted abruptly.

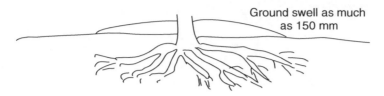

Figure 5.1 Heave of soil as result of tree removal

Drainage can also have an effect on foundations, since the location of a foundation on a drain may result in the load from the building not being distributed onto the soil, but may result in it being distributed into a void. As a result, all foundations should go below the invert level of all drains where the drains are less than 1 m from the foundation.

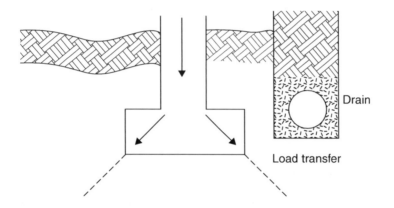

Figure 5.2 Foundations below invert level of drain

Foundations depths are also dependent on difference in level of the surrounding ground. Where the wall acts as a retaining wall against the ground, the foundation depth should be taken from the lower ground level.

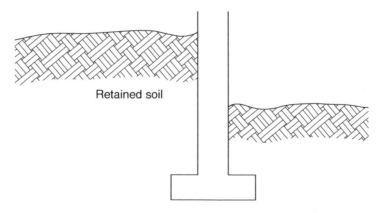

Figure 5.3 Foundation depth in relation to different ground levels

For small to medium size buildings with good ground conditions, there are two types of foundations which may be considered:

(1) Strip foundation – here the foundations are created by walls constructed below the ground using engineering bricks or concrete blocks with a weak concrete mix between the cavity;
(2) Trench foundation – here the foundations are filled with concrete to within 150 mm of the ground from where a cavity wall or solid wall construction is carried up.

For poor ground conditions, or where tree activity could be a problem, then an engineered solution would be required and expert assistance should be sought.

Foundations should be situated centrally under the wall. Approved Document A of the Building Regulations gives guidance on the dimensions of a foundation. It states that the minimum thickness 'T' of concrete foundation should be 150 mm or 'P', whichever is the greater, where 'P' is derived using the 'minimum width of strip foundation' table.

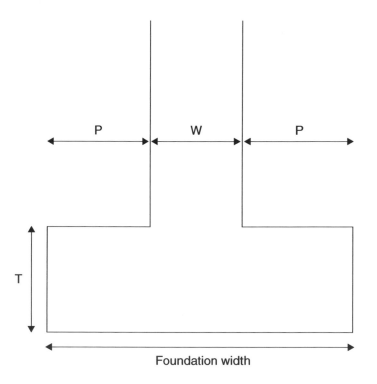

Figure 5.4 Foundation dimensions

Except where strip foundations are founded on rocks, the foundation should be 450 mm in depth to avoid the action of frost. In areas which have a long period of frost, this depth would need to be increased. As a norm, the minimum depth of foundation within England should be 1 metre. In clay soils, the foundation should be taken to a depth where anticipated ground movement would not impair the stability of the building.

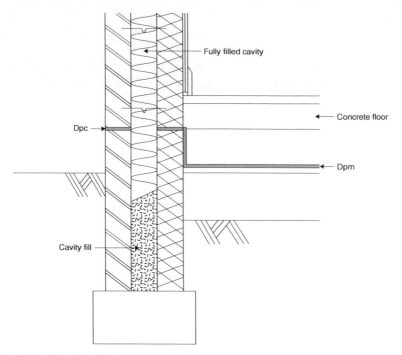

Figure 5.5 Strip foundation

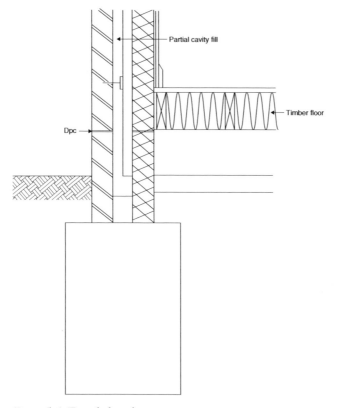

Figure 5.6 Trench foundation

Walls

A wall is designed to undertake a number of functions. These include

- Resisting the passage of moisture from the ground to the inside of the building.
- Resisting the penetration of precipitation to the inside of the building.
- Be designed and constructed so that they offer structural, fire and thermal performance.

There are various types of wall constructions. For small and medium size buildings it is normal to consider a cavity or solid wall.

Cavity walls are normally brick on the external face with a fully filled water repellent insulated product within the cavity, or a partially filled insulated product within the cavity, held by spacers against the inner wall and maintaining a clear 50 mm residual clear cavity. The thickness of the insulation material is dependent on the U-value that is required. You are advised to read the manufacturer's guidance on the appropriate type of insulation fill that would be required for your construction and location since in some regions of the UK cavity wall construction is not recommended – refer to BS 8104: 1992 Code of Practice for assessing exposure of walls to wind-driven rain.

The external face of a cavity wall is tied to the internal wall, which is normally a lightweight autoclaved aerated concrete block. The wall ties are spaced at 900 mm horizontally and 450 mm vertically for cavity widths between 50 mm and 75 mm and 750 mm horizontally and 450 mm vertically for cavity widths between 76 mm and 100 mm. At openings the wall ties are doubled up.

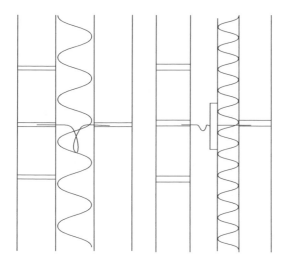

Figure 5.7 Cavity wall

Where the wall is a solid wall, and likely to be exposed to very severe conditions, the wall should be protected with a render. The render should be around 20 mm thick and have a textured finish. It is important to note that the render should not bridge the damp proof course (dpc). Solid walls are normally insulated on the inside face and have a plasterboard lining as a finish.

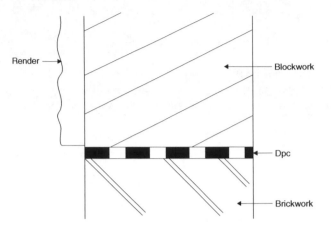

Figure 5.8 Solid wall dpc

All external and internal walls require a damp proof course . This is normally a product either of bituminous material or polyethylene. The damp proof course should be continuous along the length of the walls and should be lapped with the damp proof membrane (dpm) located within the floor.

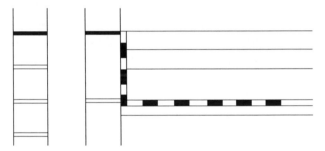

Figure 5.9 Dpc/dpm junction

It should be noted that within a cavity wall construction, the dpc should not cross the cavity. In addition when a cavity wall is proposed, the cavity should be taken 225 mm below the level of the damp proof course. The external wall damp proof course should also be located 150 mm above the level of the adjoining ground, but it is important to ensure that the internal wall damp proof course should be at finished floor level. To assist in the transfer of moisture through the external leaf, there should be weep holes every 900 mm in the external wall.

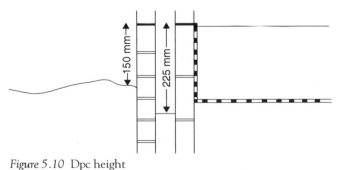

Figure 5.10 Dpc height

Floors

A floor is designed to undertake a number of functions:

- Resist the passage of ground moisture to the upper surface of the floor.
- Resist the passage of natural ground gases such as radon and methane. For more information please refer to Building Research Establishment BR414 Protective Measures for Housing on Gas Contaminated Land 2001 and BR211 Radon: Guidance on Protective Measures for New Dwellings 1999.

Floors for buildings other than those which are used to store goods, and where the floors are bearing onto the ground, should be constructed so that their structural and thermal performances are not compromised.

There are various types of floor constructions. For small and medium size buildings it is normal to consider a ground bearing slab or a suspended timber floor.

All groundbearing floors should have a base of well-compacted hardcore, no more than 600 mm in depth. The hardcore should be made from clean broken brick or other suitable material. The use of building rubble with rubbish and vegetation is not appropriate. The hardcore should have a good 50 mm sand binding prior to the installation of the damp proof membrane. After the damp proof membrane a good concrete slab is installed, insulation is placed on top of the slab and under the screed. Further guidance can be found within BS 8102 1990 Code of Practice for Protection of Structures against Water from the Ground.

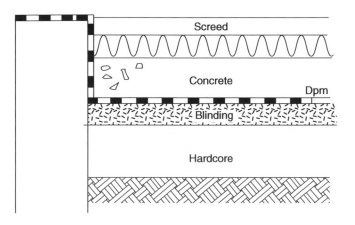

Figure 5.11 Solid floor construction

A suspended timber floor has a similar make up to the groundbearing slab, apart from the provision of a damp proof membrane and the provision of a raised timber floor. When designing such a floor, it is important to provide a ventilated air space of at least 75 mm from the ground covering to the underside of the wall plate and at least 150 mm to the underside of the floor. The void needs to be ventilated on the two opposing external walls by the use of airbricks.

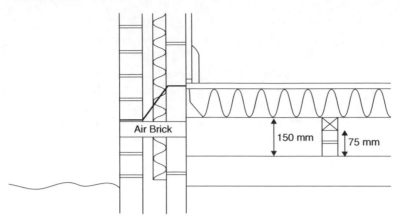

Figure 5.12 Timber floor construction

Roofs

There are two basic types of roof construction that can be used for small to medium size buildings:

- Pitch-roof construction.
- Flat roof construction.

Pitched roof

The traditional method of constructing a pitched roof is with the use of timbers to form rafters. These roofs have varying pitches but rarely go below 15° and never above 45°. 'Trussed roofs' are another form of a pitched roof and is created using trusses manufactured within a factory and delivered to site complete.

Traditional roofs are made up from rafters, which may be supported by purlins and struts, depending on the span and size of the rafters. The purlin runs at right angles to the rafters and is supported either within the gable wall or by a strut. The struts run at right angles to the purlin and need to be supported on a loadbearing wall or other structural element to help distribute the load of the roof. They are normally positioned every fourth rafter.

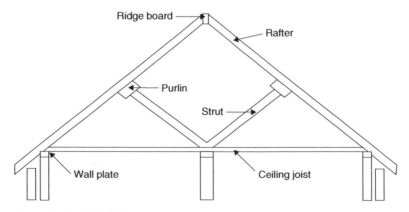

Figure 5.13 Pitch roof

The rafter size is dependent on the length from the wall plate to the ridge board as well as the type of roof covering and its weight. For guidance please refer to *Span Tables – 2nd Edition* published by TRADA Technology. Typically, the spacing between the rafters should be 600 mm centres, the rafters should be nailed to the wall plate, which is bedded onto the top of the inner wall. To enable the rafter to connect fully with the wall plate a triangle section called a 'birdsmouth joint' should be created. The provision of vertical restraint straps should be provided at every 2 m centres to tie the wall plate to the wall and to prevent uplift of the roof. The provision of horizontal restraint straps should be used to tie the gable wall to at least three rafters.

Ceiling joists are sometimes connected to the rafters at the wall junctions and are sometimes supported by binders that run at right angles to the ceiling joist and are supported to the purlin by timber hangers. Supports for the ceiling joist are dependent on the span.

Ventilation to the roof is important to ensure that any condensation is removed and, as such, a vent of 10 mm at the eaves and a 5 mm vent at the top of the pitch are required for roofs with a pitch more than 15°. Where the pitch is less than 15° the vents at the eaves should be increased to 25 mm.

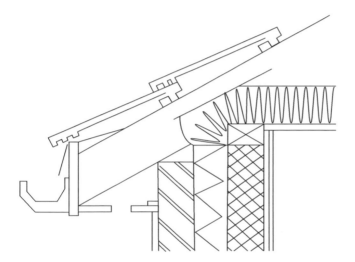

Figure 5.14 Eave junction

For more complex roof layouts, it would be advisable to seek the services of a building engineer for structural guidance.

Flat roof

Spanning joists between two walls creates this type of roof. The size of the joist is dependent on the span required. For guidance please refer to *Span Tables – 2nd Edition* published by TRADA Technology.

The joists are spaced 450 mm centres normally and have a 'firring piece' – this is a tapered piece of wood fixed at the top of the joist running the full length of the joist to a depth of 50 mm at the gutter. Alternative the joist could be laid to fall. This assists in drainage of rainwater from the roof.

Depending on the span of the roof, horizontal struts between the joists should be provided. In addition, the roof should be tied to the wall using horizontal restraining straps fitted across at least three joists and firmly connected to the wall to prevent uplift from high winds. Alternatively, vertical restraint straps can be fixed, with a minimum of one joist per 2 m run.

The deck to the flat roof should be exterior quality plywood of at least 19 mm thick or roofing grade boards stamped with WBP (weather water boil proof). The board should be securely fixed to the joist paying special attention to the edges. On top of this deck a proprietary insulation product should be installed to offer adequate thermal insulation. This type of flat roof is called a 'warm deck' flat roof. This type of construction is by far the best method since there is no need to fill the void between the joists with glass fibre insulation and as a result no need to ventilate the void or provide soffit or eave vents.

The thickness of the thermal insulation is dependent on the product type that you propose and the U-value required under the Building Regulations. I would recommend that you contact a building engineer, who can advise on the required thickness of roof insulation and products suitable for your proposal. The insulation product would require waterproofing and this is traditionally achieved by using three layers of felt laid onto hot bitumen and finished off with 12.5 mm spar-stone chippings covering the whole felt. At the abutment of the wall, the felt is taken up and tucked into a raked-out mortar joint and then pointed in. This detail is finished off with Code 4 Lead flashing.

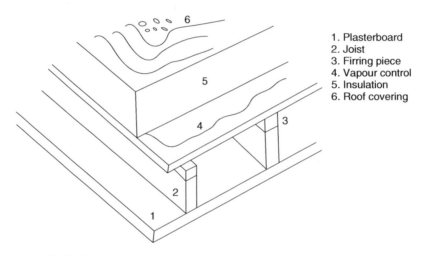

1. Plasterboard
2. Joist
3. Firring piece
4. Vapour control
5. Insulation
6. Roof covering

Figure 5.15 Flat roof

Drainage

There are different types of drainage, which may be appropriate for use within and around your building.

- Foul drainage – This conveys the flow of foul water from sanitary appliances or from water used for food preparation, cooking or washing from within the building to either a public sewer, private drain, septic tank or other appropriate discharge.
- Public sewer – Where a drain serves more than one building, it is considered a public sewer and under the Water Industry Act 1991 the owner or occupier of the building has a right to connect to the public sewer. There are some restrictions and you should contact your local sewer undertaker for their guidance.
- Surface water drainage – This conveys rainwater from roofs or paved areas to either a soakaway or private drain/public sewer.

- Soakaway – is an infiltration drainage system. For more information on the design of infiltration drainage system, I refer you to CIRIA Report 156 – Infiltration Drainage – Manual of Good Practice. Ideally, soakways should not be built within 5 m of a building or road or in area of unstable land. A percolation test should be carried out to determine the capacity of the soil. Soakways should be designed for the total amount of rainwater that is expected. They are generally formed from square or circular pits filled with rubble.
- Combined drainage – In some circumstances the drains may carry both foul and rainwater into the public sewer.

All drainage layouts should be kept simple and where possible laid in straight lines, but slight curves are possible. Acute changes of direction should be kept to a minimum. Drainage pipes should be laid to an even gradient and bedded in granular material like pea shingle. UPVC pipes should not be encased in concrete.

There should be sufficient and suitable access points to enable clearance of blockage. There are four types of access points.

1. Rodding eyes – capped extensions of the pipes.
2. Access fittings – small chambers on pipes but not with an open channel.
3. Inspection chamber – chambers with working space at ground level.
4. Manhole – 1 m+ deep chamber with working space at drain level.

Outlets

All waste water should connect to a foul or combined drainage system. They should not connect into the rainwater drainage. Sanitary waste pipework should discharge into a soil and vent pipe (SVP) or a stub stack and then connect to a drainage system. Waste water can either connect to the SVP, stub stack or a robbable gully at ground floor level.

Stub stacks can be used if they connect into a drain run which is ventilated. No connection should be made to a toilet located less than 1.3 m from the bottom of the drain.

Soil and vent pipes (SVPs) are needed to ventilated and drainage system and prevent lost of pressure within the waste traps. When taking waste from a toilet, the minimum size should be 100 mm diameter. The SVPs discharge directly to external air and finishes at least 900 mm above any opening into the building nearer than 3 m.

Rodding access should be given at the base of all SVPs and at the change of direction of all waste pipes. This ensures that provision of offered to minimise the risk of blockage.

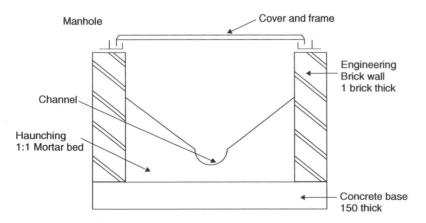

Figure 5.16 Manhole

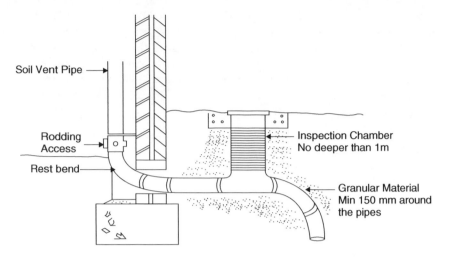

Figure 5.17 Foul water drainage

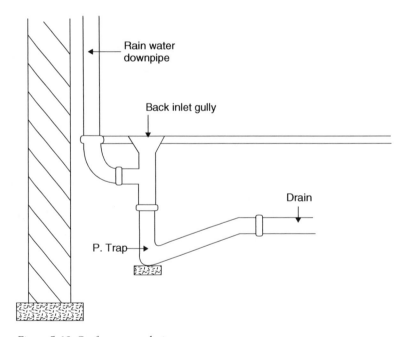

Figure 5.18 Surface water drainage

6 Pre-contract

This chapter discusses matters that typically need to be considered (aside from the detailed design of the project) prior to commencing minor works projects.

Many small builders and their customers will inevitably hold the opinion that, for minor works, detailed specifications represent just so much padding, an unnecessary cost burden which could have been passed over with no tangible detriment to the project. That is undoubtedly so in many cases. However, there are too many instances where the use of even a simple form of specification would have prevented errors, delays and costs from being incurred, and there are few projects that would not have benefited from having an appropriate specification to which all parties could refer, and thus potentially save time and costs overall.

This chapter will therefore consider the reasons and benefits for having a specification; what can go wrong when a specification does not exist for a project; and what actually constitutes a specification. The chapter also looks at standard forms of contract and the responsibilities of each of the parties to the contract. Finally, the chapter discusses the Construction (Design and Management) Regulations.

6.1 Why have a specification?

Some years ago, one of the many well-known professional journals serving the construction industry illustrated most succinctly the answer to the question that is posed by the main title of this section, in so doing adopting the age-old premise that a picture paints a thousand words. Readers (well, older ones anyway) may recall that the journal in question published a series of cartoon images depicting 'what the client wanted' – this being a simple child's swing comprising an old car tyre suspended on a length of rope from the conveniently-located horizontal branch of a tree. The cartoon then proceeded to illustrate various other forms of the swing, including 'what the architect suggested', 'what the project manager approved', 'as designed by the engineer', 'what the contractor actually installed', 'what the health and safety inspector insisted upon', and so on. The cartoons drew many a wry smile in professional design offices around the land – as they say, the truth is often funnier than fiction. (For any reader curious to see a version of the cartoons, a simple internet search for 'what the client wanted' should lead to numerous examples of the genre.)

Nonetheless, the point is well made – without a well-written specification, how will the client ensure that the final product matches the desired performance criteria and/or aesthetic appearance?

6.1.1 *Reasons and philosophy behind the specification*

Perhaps one of the oldest recorded specifications may be found in the Bible in the book of Genesis, Chapter 8, verses 14–16, wherein the 'client' (God) instructed the 'contractor' (Noah) as follows:

> Make thee an Ark of Gopher wood; rooms shalt thou make in the Ark and shalt pitch it within and without with pitch . . . And this is the fashion which thou shalt make it of: The length of the Ark shall be three hundred cubits, the breadth of it fifty cubits, and the height of it thirty cubits . . . A window shalt thou make to the Ark, and in a cubit shalt thou finish it above; and the door of the Ark shalt thou set in the side thereof; with lower, second and third stories shalt thou make it.

As clear, concise and straightforward a specification as one could wish for. In one simple document, the client has advised the contractor regarding his choice of material, details of the internal and external damp-proofing provisions, the dimensional limitations within which the project is to be constructed – including number of stories for accommodation purposes – and the nature and location of the means of access and size of opening window for natural lighting/ventilation. Depending upon the nature and depth of one's individual beliefs, one can only speculate what the outcome for the future of mankind might have been, had the contractor been left to his own devices.

Fortunately, in this instance it appears that the client was also able to influence the prevailing atmospheric conditions during the contract period and so there was apparently no need to include any complicated contract clauses to deal with circumstances where the contractor might have made any claims for delays due to poor weather.

So, a specification is a description in detail of the building enterprise. It is considered by most professionals to be one of the most important documents in, and forms part of, the building contract, for it is customary to include in it references to the conditions under which the work is to be executed.

The specification for minor works projects is most commonly prepared by the project designer – possibly a building engineer, an architect or an architectural technologist, for example. Only on larger projects, where the budget may be less restrictive, is it common to find that a specialist quantity surveyor is employed to assist in writing the specification or in producing detailed bills of quantities for the works. In some cases, the bills of quantities effectively may form a large part of the specification.

6.1.2 *The pitfalls . . . and the benefits*

In order to take advantage of the benefits of having a specification and to also avoid falling foul of the potential pitfalls that can afflict any building project, it is important that the clauses in a specification must be absolutely precise and definite, so that no dispute can arise over interpretation.

The project designer should ideally never insert vague phrases and words. Expressions such as 'in a proper manner', 'all complete', 'as required' or 'the best materials' should be particularly avoided. If the word 'best' is used, it should be defined and never be inserted unless intended.

The specification should not demand impossibilities, such as 'boarding free from *all* knots'.

It is essential that all information be complete and accurate. If any item is unknown at the time of writing the specification, an appropriate provisional sum must be inserted, since once the agreement is signed, the builder is liable only for work and materials actually specified and any additional expense incurred is charged as extra.

Brevity, by reason, decreases opportunities for misunderstanding; but clarity must never be sacrificed. Tabulation wherever possible, makes a specification easier to read and repetition should be avoided. If any part or member of a building is duplicated, a full description need be given only once unless, for example, a sub-contractor, who would require full information, is employed.

Ideally, each item should be dealt with under its own trade heading, the main trade headings varying somewhat with the nature of the job. Each trade should commence with specifications on the quality of materials used.

As an example, the specification in relation to concrete works should either identify mix proportions of sand, aggregate and cement and water content (if on-site mixing of small quantities is to be employed) or, alternatively, it may be more appropriate to specify a strength grade classification if the concrete is to be sourced from a ready-mix supplier.

The specification may also identify details of any sampling and testing requirements for the concrete to ensure consistent quality, strength and durability or possibly there may be a requirement to prepare a test panel to a particular level of surface finish quality for approval by the client, project designer or planning officer before proceeding to construct the remainder of the project.

Increasingly in the twenty-first century, specifications will refer to (for example) British Standards as a means of controlling quality of materials and workmanship. It is important, however, that the person writing the specification does not simply rely upon quoting a particular British Standard reference number, but rather that they fully understand what will result from their use of that BS reference in the specification.

As an example, a specification may refer to BS5606:1990 – 'Guide to accuracy in building' as the standard to which the builder must comply in achieving an accurately-built structure. This British Standard sets out – as its title implies – various parameters for what is regarded to be an acceptable standard of accuracy of construction that should be achievable by a competent tradesman operating in normal working conditions. So, an inexperienced person writing a specification for the erection of a wall that is to be straight and vertical might simply refer to the need to comply with this British Standard and expect that the resultant wall will be 'perfectly' straight and plumb.

What should be appreciated, however, is that compliance with the basic provisions of this particular British Standard would permit a wall of brickwork masonry construction to be as much as ± 7 mm out of horizontal alignment in any 5 m length of wall, or as much as ± 10 mm out of plumb over a 2 m vertical section of wall and still meet the minimum required standard of accuracy.

Now, there will be many instances where such deviation from a true alignment would not be problematic; however, if pre-fabricated elements are required to be subsequently installed within that part of the building, there is a possibility that they may not fit properly into the space allocated unless the specification is purposely written to identify to the builder that in this particular area the building tolerance requirements are more onerous that the basic provisions of the British Standard. Therefore those requirements for an enhanced degree of accuracy in the finished construction should be clearly identified in the specification.

6.2 What form of contract is appropriate?

6.2.1 *The importance of using a written contract*

It is disappointing that even in these enlightened times so many small building projects are commenced purely on the basis of a verbal agreement between two total strangers. Worse still, significant sums of money change hands, often in advance of the works commencing and without any real chance of recovery by the client should things go awry, except via lengthy and expensive court procedure – and sometimes not even then.

Even between two individuals or groups who are well acquainted, to proceed without a written contract is simply a recipe for potential disaster, and perhaps the end of a good working partnership or, worst, the end of a long friendship. Many private clients believe that a written quotation from a builder is all the protection that is required for the project to proceed without any potential problems, but sadly this is not always the case.

This may appear to be a very jaundiced viewpoint as there are a great many building projects undertaken and successfully completed each year in the UK by competent, conscientious builders without anything more on paper than a written quotation to the client in the form of a single page on the company letterhead. Our intention is not to malign those builders who do a fine job of work. However, if (some might say 'when') a problem occurs part-way through the project, irrespective of who has caused the problem, there are no laid-down rules with which to guide the two parties in resolving the problem quickly, economically and, hopefully, without undue rancour. Such situations do occur, even when using the best, most conscientious builders.

The construction industry has always been plagued with a small but significant proportion of builders who might perhaps be termed 'cowboys'. The popular media would perhaps have us believe that there are an awful lot of these 'cowboys' out there. In truth, however, they represent only a small minority, and it is simply that with increasing access to information we are all becoming ever more aware of their presence. Returning, then, to the theme of the opening paragraph of this section, it is even more surprising to find a significant number of projects being commenced without a formal contract.

A good example of why reliance upon a written 'quotation' can be unwise is the case of a domestic client who in the early 2000s instructed a builder to carry out various works to his home. The works comprised the erection of a two-storey side extension, with a single-storey offshoot to the rear of it, plus extensive landscaping works to the front and rear of the property. The builder had submitted a two-page letter on the company paper, which on page one carried the bold title 'Quotation for Extensions and External Works at . . . (address of property . . .)'. The letter set out in broad detail the extent of the works and, at the bottom of page two, concluded with the wording 'Estimated Price: £41,500'.

To cut a long story short, the whole enterprise ground to a halt when, with about £5,000-worth of external works still to be completed, the builder presented the client with an *interim* invoice for a little over £56,000! Not surprisingly, the client refused to pay this amount – especially as he held the opinion that the builder was trying to recover all of his abortive costs incurred by using inadequate sub-contracted bricklayers, resulting in a large portion of one elevation of the extension having to be taken down and fully rebuilt, twice!. The matter passed almost inevitably into litigation. The whole scenario was not assisted by the fact that during the course of the project the client had verbally requested several changes to the original specification (including some extra work elements, but also removing others that the client decided to undertake himself) without agreeing with the builder the cost of these variations to the original contract, let alone confirming the nature of those variations in writing.

The matter was eventually determined by a court, with neither party coming away from the courtroom financially unscathed. However, perhaps the most salutary lesson here relates to the legal costs for the case, which amounted to a significant proportion of the (inflated) contract total. These legal costs were in part a reflection of the fact that our colleagues from the legal profession spent some considerable time arguing over whether the contract amount stated in the original letter from the builder was in fact a firm 'quotation', as the client thought, or an 'estimated cost', as the builder's legal advisers argued.

One can only speculate how much unnecessary cost (to say nothing of the delays, worry and stress for both parties) might have been avoided had the work been tied up by a straight-forward formal contract. Also, one only wishes that the foregoing tale of woe was an exception that proved the rule – alas, it is only one of many similar tales that abound in this industry.

6.2.2 *Examining the different types of standard forms of contract available*

For the inexperienced professional, or for the lay client, discovering and understanding the relevance of the wide range of different types of formal construction contracts that are available in the marketplace can be confusing. Also, the contracts are, of course, legal documents and hence their wording can be daunting to untutored eyes.

It is not the intention in this book to go into great detail about the different types of contract that are available – there are many other specialist published works that fulfil that requirement. However, as in all cases when dealing with legal documents, when in doubt, the watchword is to take sound professional advice.

There are three main groups of contract types, which relate to the three main methods of procurement for building projects, these being:

- Traditional procurement
- Design and build procurement
- Management procurement.

The client's ultimate choice of procurement method and hence of contract type will be subject to a number of factors, including the type and size of project being considered; how much control the client intends to have either personally or through the engagement of a consultant supervisor; the degree of flexibility that the client can comfortably accommodate in regards to the final project cost; any restrictive matters, such as will be discussed elsewhere in this book; and the likelihood that significant variations will occur during the construction phase of the project.

6.2.2.1 *Traditional procurement*

This has been the building industry's 'standard' form of building procurement going back several generations and it is also arguably the most appropriate form of procurement for the majority of minor works projects. The principal point to understand with this method is that the process of designing the project is kept separate from the construction phase. Also, full documentation should be available before the builders are invited to tender for the project. For traditional procurement projects, there are three main types of contract, these being:

- *Lump sum contracts* – where the contract sum is determined prior to work commencing.
- *Measurement contracts* – where the contract sum is not determined until after the works are completed, usually by re-measurement of the works.
- *Cost-reimbursement contracts* – where the contract sum is determined by an ongoing assessment of the actual costs of labour, plant and materials.

6.2.2.2 Design and build procurement

As the title implies, with this method of procurement the builder is responsible not only for the construction of the project but also for the design process. The potential disadvantage here is that the client may lose a degree of control over the detailed design and so, for many domestic building projects, this procurement method is often regarded as unsuitable. For design and build procurement projects, there are again, three main types of contract, these being:

- *Turnkey contracts* – where the client chooses a complete package, most commonly to a standard specification package from a commercial firm.
- *Design and build contracts* – where the documents also set out the contractor's design obligations.
- *Design of specific elements contracts* – which is similar to the above, but obviously only relates to specific parts of the project.

6.2.2.3 Management procurement

In this method, the contractor is responsible for managing the works and for letting and supervising subcontractors to carry out the works. The 'management contractor' may be a building firm or perhaps a firm of professional consulting project managers.

This method of procurement is rarely used on minor works projects, as the overheads generated cannot usually be justified in terms of overall project savings.

6.2.3 The Joint Contracts Tribunal

By far the most widely used and well-known standard construction contract forms are those prepared and published by the Joint Contracts Tribunal Limited (JCT), which was established in 1931 to publish and maintain by amendment, as and when necessary, a standard form of contract for the construction industry.

The JCT membership originally comprised the Royal Institute of British Architects and the National Federation of Building Trades Employers (NFBTE), however its current membership includes such bodies as the British Property Federation, the Local Government Association and the Royal Institution of Chartered Surveyors, as well as the Royal Institute of British Architects. Since 1931, when the original standard building contract was drawn up, the JCT has produced a further five editions, culminating in the present suite of documents which became available from 2005.

There are currently a full suite of contract forms available from the JCT covering all potential needs and projects. For minor works projects such as those that are intended to be the subject of this book, there are two principal contract ranges that apply, these being:

- Minor Works Building Contracts
- Home Owner Contracts.

There are several individual contract forms within both these groups, the primary examples of which are discussed in more detail below:

6.2.3.1 *Minor Works Building Contract (MW)*

This is the most commonly used form of contract for minor building works and accordingly, the advice given in Chapters 7 and 8 of this book is based primarily upon the assumption that this form of contract is likely to be used.

The MW form of contract is appropriate for projects where the proposed work is not complex in nature, where the scheme has been either designed by the client or by a consultant on the client's behalf and where drawings and a specification and/or a schedule of work will be provided to the contractor to determine the extent and quality of the finished works. There is also an assumption that this type of contract will be administered on the client's behalf by an appropriate professional representative. On acceptance of the successful tender, the documents provided to the contractor become the contract documents and thus define the works on which the contract sum is based.

The contract sum is intended to be a lump sum under this form of contract; however, it provides an option for limited fluctuations in the contractor's quoted prices where such fluctuations might arise from, say, government-driven contribution, levy or tax changes and the like.

The MW form of contract is not intended for use in cases where detailed bills of quantities are required or where detailed control procedures are necessary. Most significantly, it is not suitable for contracts where the contractor is expected to be responsible for the design of all or part of the works.

6.2.3.2 *Minor Works Building Contract with contractor's design (MWD)*

This contract form is very similar to the MW form of contract described above, but in addition it provides for those instances where the contractor is expected to design all or part of the project works.

6.2.3.3 *Building contract for a home owner/occupier who has not appointed a consultant to oversee the work (HO/B)*

The JCT Home Owner contracts are, unlike the Minor Works series, intended to be consumer contracts for use by residential occupiers. Because they are intended to be used by 'lay persons' they are written in clear, simple non-technical language, as required by the Unfair Terms in Consumer Contracts Regulations 1999. They are appropriate for use on small domestic building work such as extensions and alterations and where the work to be undertaken is to be carried out for an agreed lump sum. It is intrinsic to these forms that the works should not be complex in nature.

The contract will be based upon scheme drawings and/or a specification. The lump sum contract price will normally be inclusive of Value Added Tax and the basic form of the contract provides for a single payment on completion of the works; however, payment may be made by instalments by agreement between the parties.

This form of contract is intended to be administered by the client directly with the contractor and there is no provision within the contract for the involvement of an independent administrator on the client's behalf.

6.2.3.4 *Building contract for a home owner/occupier who has appointed a consultant to oversee the work (HO/C)*

This contract is very similar to the HO/C form of contract discussed above, except that it provides for the appointment of a contract administrator on behalf of the client.

6.2.3.5 *Home Repair and Maintenance Contract (HO/RM)*

This is also a consumer-type contract for use by a residential occupier. It is, as its name implies, intended for projects to carry out small-scale repairs and maintenance of a straightforward nature to domestic properties. It is not intended that the client will have appointed an independent consultant to administer the contract.

The contract price may be a lump sum or, alternatively, it could be based, if appropriate, upon a rate per hour/day, plus materials at cost plus a handling charge to the contractor. Again, it is normally the case that the quoted price will be inclusive of Value Added Tax. Most significantly, this form of contract *only* allows for payment on completion of the works and so it should only be used where the works are intended to be of very short duration and certainly not longer than 20 working days.

6.3 Contract documents

As suggested in the previous section, the most commonly used form of contract for minor building works is the JCT Minor Works (MW) form of contract and so the details given in this section of this chapter tend to concentrate upon the level of detail that would customarily be included in the contract documents for this form of contract. This should enable a degree of consistency and cross-referencing of the points raised in this chapter with those in Chapters 7 and 8 of the book.

6.3.1 *Design drawings and details*

There is an ancient proverb which says that 'a picture paints a thousand words'. Accordingly, one of the most important elements of the contract documents is the suite of drawings that describe the project. The size, number and complexity of the contract drawings will normally be a reflection of the size and complexity of the project. For the purposes of illustration, some examples are worth considering.

A project for, say, the replacement of the roof covering to a multi-storey building with a flat roof might be adequately depicted with a single layout plan showing the building roof, graphically identifying the extent of the roof area to be re-covered. The principal quantities (total area, lengths of kerbs/upstands, lengths of copings, number of rainwater pipe penetrations, etc.) could also be stated on the layout drawing, or they could be set out separately in the contract documents. It would be left to the contractor tendering for the project to choose a suitable product for the roof covering, which should, however, be specified to have certain performance characteristics, perhaps including a minimum guaranteed service life.

By comparison, most domestic extension projects are undertaken using the same drawing that was submitted for approval under the Town and Country Planning Act and the Building Regulations. In order to gain Building Regulations approval, the drawing would normally include a moderately detailed specification, which might state details of particular manufacturers' products.

Because the majority of domestic extension designs are not produced by established architectural practices, the quality of the drawings produced varies widely. It is a sad observation that in far too many cases the design of domestic projects (and indeed also for some larger, non-domestic projects) is awarded to the lowest bidder rather than necessarily to the most capable designer. There are numerous examples each year of project drawings being approved by Building Control officers which, whilst they include sufficient information to gain Building Regulations approval, are plainly inadequate to permit accurate, reliable construction of the intended project on site. As the renowned nineteenth-century writer John Ruskin is reported to have said: 'There is hardly anything in the world that some man cannot make a little worse and sell a little cheaper, and the people who consider price only are this man's lawful prey . . .'

This is important, because the more detail that is shown on the drawings, the less risk there is of the contractor misinterpreting the designer's intent and/or the client's wishes. It should be remembered that if the contractor submits a competitive quotation for 'works as outlined on drawing number XXXX', then that is precisely what has been included in the quote – unless the contract documents say he should have done otherwise. If during the construction phase the work content is increased due to an omission of a detail by the designer, but which is demanded by, say, the Building Control officer, the contractor will be likely be entitled to an additional payment (see section 6.3.4 below). One element of a recent legal case for a domestic extension project turned around the fact that the builder, in preparing his quotation, had clearly stated that he had included for the works 'as depicted on the project drawing'. When subsequently the Building Control officer demanded the inclusion of additional fire protection measures (which had not been clearly shown on the drawing) the builder requested an additional payment, which the owner contested on the grounds that he expected the builder to build an extension which would be granted a Completion Certificate by the Building Control officer. The builder was successful in securing an additional payment for this work on the grounds that had it been shown on the drawings in the first place as being a regulatory requirement, the builder would have had the opportunity to include the cost of the fire protection in his original quotation.

By comparison, the architectural drawings for more complex projects might perhaps include:

- A general arrangement/layout plan.
- Detailed cross-sections showing small-scale constructional details.
- Enlarged details of the principal constructional elements.

In addition, other elements of the works may warrant their own drawings and/or details in addition to the architectural design drawings, and may include for example:

- Structural details (foundations, structural steelwork connections, reinforced concrete details, etc.).
- Details of service layouts (electrical wiring and fittings, plumbing and heating, drainage details including inspection chamber construction, etc.).

- Site restrictions affecting how the contractor may carry out the works, especially any that may affect neighbouring landowners (see also Chapter 4).
- Prescribed locations for site compounds, etc.

The above list is not intended to be exhaustive. . . .

The 'golden' rule when producing the contract drawings is for them to be as detailed and complete as is reasonably practicable. Never leave anything to chance and never assume that the contractor will necessarily read the mind of the designer and/or the client when carrying out the works on site. The contract drawings – in conjunction with the works specification – should ideally direct the contractor clearly and concisely toward the client's desired end-product (the reader is referred to the opening comments at section 6.1 above).

6.3.2 *Start date/contract period*

There is an old adage which goes 'more haste, less speed', and for some perverse reason this certainly seems to apply to construction projects. There are numerous horror-stories of projects where either the design, or the tender, or the construction phase(s) have been rushed through to comply with some deadline or other, with the end result being either poor workmanship, escalating costs, legal problems, over-runs on completion, or even a combination of these issues.

The clever trick is to think ahead, of course. By all means set challenging target dates but never, ever make those target dates unreasonable for the prevailing circumstances. It the programme is not reasonable, then a court might be reluctant to support a claim against a contractor who failed to meet them – even if he 'agreed' to them in the first instance by accepting the contract.

The effect of target dates upon health and safety matters is also very important. For example, the Construction (Design and Management) Regulations 2007 (see section 4.6 below) demands that the client should allow adequate time between awarding the contract and starting on site for the work to be properly planned by the principal contractor.

Similarly, the contract period should be practicable and reasonable for precisely the same reasons. The JCT MW form of contract makes provisions for the assessment of 'liquidated damages' in circumstances, for example, where the contractor perhaps may fail to meet a completion date. However, excluding factors relating to reasonable delays due to bad weather and the like, an unreasonable contract period would be unlikely to make a good basis for a liquidated damages claim.

Also, make allowances in the project programme, especially at the beginning of the project, for any neighbour-related issues (see Chapter 4) to be resolved. As an example, if a notice under, say, the Party Wall etc. Act 1996 had to be served upon an adjoining property owner for proposed works to an existing party wall, the notice period required by the Act is two calendar months. If the adjoining owner dissents to the notice, the resulting dispute procedure might be completed within that timeframe, but the commencement of the notifiable works could only commence earlier than the end of the two months notice period with the express written agreement of the adjoining owner, and in many cases that may not be forthcoming.

Similarly, where the works will involve say, the erection of a scaffold platform on land belonging to an adjoining property owner, if a scaffold licence has been drawn up to facilitate this, it is important that the project programme accurately reflects any time limits that may apply for the removal of the scaffold from the third-party land.

6.3.3 Insurances

One of the most important factors in ensuring an appropriate apportionment of the level of risk between the client, the designer and the contractor is the choice of the correct form of contract for the project.

That factor aside, however, it is important to ensure that an appropriate degree of insurance cover is demanded by the works specification for the project. Such insurance cover may take several forms, including:

- Third-party insurance
- Employer's liability insurance
- Professional indemnity insurance
- Collateral warranty
- Product and/or installation warranty.

6.3.3.1 Third-party insurance

Sometimes referred to a 'public liability insurance', such cover is a basic requirement for all companies, whether contractors, suppliers, transport providers or construction firms. The policy provides cover in the event that the company or its employees cause accidental damage to third-party property (i.e. property owned by someone other than the insured company) or to other persons not directly employed by or part of the insured company.

The level of financial cover required to be provided for any specific contract will vary, dependant upon the project being undertaken, and hence this will be related to the likely degree of risk. For example, a project to demolish a reinforced concrete frame structure in close proximity to premises in separate ownership will probably require a higher level of financial cover than will, say, a project to carry out shallow excavations on a green-field site well away from any third-party boundaries/buildings. Projects that will involve potentially hazardous materials will also commonly require much higher levels of third-party cover.

The cost to the contractor/designer of obtaining third-party insurance will also vary dependant upon the type of work undertaken by the insured party and their past claims record. Inevitably, these insurance costs are passed on to the client within the overall fee/works costs for the project and so it is important to specify a level of third-party cover that is appropriate for the project and that is not unnecessarily high.

6.3.3.2 Employer's liability insurance

It is a legal requirement in the United Kingdom that all employers provide a minimum level of insurance cover to protect their employees against the consequences of serious injury or death whilst at their place of work. The level of cover required does increase over time but in the early twenty-first century the minimum level of cover is generally taken to be of the order of £10 million for each and every claim.

The works specification should demand that evidence of this cover be provided by all consultants and contractors before commencing work on the project.

6.3.3.3 Professional indemnity insurance

Professional indemnity insurance (PII) provides protection to the client against the possibility that the designer/consultant may act negligently in performing his or her required duties under the contract. Again, the level of PII cover required will vary dependant upon the level of financial risk involved. For example, a consultant contracted to design the foundations for even a small structure would need to have a level of cover commensurate with the potential full cost of demolishing and rebuilding the structure if the foundation design was subsequently found to be inadequate – even though the actual cost to construct the foundations might have been only a small fraction of the overall build cost for the project.

It is a requirement of membership of all professional bodies that their corporate and chartered members who are engaged in private practice work should ensure that they are protected by adequate levels of PII cover for the works they undertake. It is not only firms of dedicated consultants that need PII cover; a contractor engaged in, for instance, design and build works would need to carry appropriate PII cover for the design elements.

Most professional bodies will currently specificy that the minimum required level of cover should be not less than £500,000 for each claim, but as indicated above, the actual level of cover held by each consultant/designer should be appropriate to the service and project they are undertaking.

Unlike most other types of insurance cover, any claim under a PII policy would be relevant to the cover held at the time of the claim and not the cover in force at the time the alleged act of negligence took place – which may of course have been several years prior to the emergence of the problem. Thus, PII cover needs to be continuously maintained long after the completion of a project and individual self-employed designers/consultants seeking to retire need to carry some form of reducing PII policy (sometimes referred to as 'run-out' cover) for a number of years following the completion of their final projects.

6.3.3.4 Collateral warranty

A collateral warranty is commonly a requirement of a financial institution which is providing the funds for medium to large construction projects. It is a joint agreement between the client, the finance provider and the designer or contractor and ensures that, for example, a project designer would be contractually obliged to maintain an adequate level of PII cover for the project for a fixed period of time following completion of the project – this period may be 10 or 12 years or possibly longer in certain circumstances.

Such warranty documents are complex and legally binding and should be prepared with the assistance of an appropriate professional – possibly a solicitor, although most specialist insurance brokers will also be able to provide such assistance.

6.3.3.5 Product and/or installation warranty

We are all familiar with product warranties for day-to-day items: most household electrical appliances will, for example, carry a 12-months guarantee against failure and many family cars currently offer a warranty against serious mechanical failure for three years, with some manufacturers now offering up to seven-year warranties on their products. Many recent improvements in this regard have come about as a result of formal legislation, including the Sale of Goods Act 1979, the Supply of Goods and Services Act 1982, the Sale and Supply of Goods Act 1994 and the Sale and Supply of Goods to Customers Regulations 2002, etc.

Similarly, a specialist contractor providing a service or, more commonly, installing a product will be expected to provide a guarantee against the failure of the installation or product. In the past, these 'guarantees' have commonly been in the form of a simple certificate printed on the company headed paper. The potential problem with this type of warranty is that the 'guarantee' is valid (i.e. the client would be able to instigate a claim under it) only whist the company remains in business and, indeed, there have been numerous instances where a smaller company faced with a large claim decides to enter into voluntary liquidation to protect its shareholders/directors, thus leaving the customer with no protection at all.

It is commonly a requirement, therefore, that any such 'guarantee' should be supported by an appropriate policy from a third-party insurer, thus maintaining a degree of financial protection to the client even if the installer/supplier of the product were to cease trading for any reason. This could be written into the works specification for a project where specialist services or products are required. Such services/products might include, for example:

- Specialist single-ply roof coverings
- Damp-proof course treatments
- Tanking membranes to basements
- Corrosion protection to steelwork
- Timber treatments against fungal decay or wood-boring insect attack
- Double-glazing units
- Mechanical/electrical plant installations
- Underpinning works.

(This list is not intended to be exhaustive.)

The warranty period will vary according to the circumstances and the product/service in question; however, most of the typical examples shown in the above list would be expected to carry an insurance-backed warranty or guarantee for a period of between, say, 5 years and 30 years – or even more in some instances.

6.3.4 *The tender process*

Having just read the title of this section, one needs to be assured that this is not intended to be the title of the latest romantic novel from a well-known publishing house, nor is it a method of ensuring that one's favourite cut of steak is not as tough as old shoe leather. Rather, it is the process by which a client may seek submission of quotations from consultants or contractors for the design, management and construction of their proposed project.

The word 'tender' in fact derives form the old French *tendre* (to offer) and in the context of this chapter has the appropriate dictionary description of 'a written offer to contract goods or services at a specified cost or rate; a bid'.

Many larger organisations have a policy to be transparent in their dealings, and especially in the manner in which they protect themselves, their clients and the contractors submitting tenders against possible fraud. Thus, it is common practice to dictate that tenders should be submitted to the intended destination office in an envelope clearly marked to indicate that the contents should not be opened until a pre-set date and time. This process is commonly undertaken by two persons, each of whom 'witnesses' the actions of the other, and is clearly intended to help prevent a situation occurring where the contents of a tender (i.e. the quoted price) submitted and opened in advance of the due date/time might be transmitted to one or more of the other prospective contractors who would thereby gain an unfair advantage.

The actual tender submission would at this early stage probably comprise a single sheet of paper (the 'tender form') which should state the name of the project, together with any appropriate client reference number for the project. It should, additionally, have a blank space where the contractor may write in the amount of the tender (either including or excluding Value Added Tax, as preferred by the client).

The tender form should carry legally-binding statements by which the contractor agrees to be bound if successful in his bid. Such statements might include, for example:

> We the undersigned, do hereby tender and undertake to perform, execute and do all works, provide all materials within the time or times described, mentioned or implied in the Schedules, Specifications and Drawings to the entire satisfaction of [the client].

> We understand and accept that the works shall be executed in accordance with the current edition of the [name of the particular contract – e.g. JCT Agreement for Minor Works 2005 Edition, or similar] and, if required, we undertake to sign this agreement. We also understand that if a formal Agreement is not executed, a written acceptance of the tender shall constitute a binding contract.

> We will furnish you, if so requested, with a fully priced copy of the Schedule in support of our tender and agree that the prices therein shall rule in any variation to the works. [Please see also to the following section of this chapter, 'Provision for variations'.]

> We agree to execute dayworks, for which authority shall be given by the client's representative, at the rate of____% profit on the basic cost of labour and____% profit on the basic cost of materials. The rates tendered shall be inclusive of all overheads, profits and all other associated charges.

Finally, the tender form should provide a space for the submitting contractor's representative (usually a Director or senior manager) to sign and date the document and to insert the name and address of the company submitting the tender.

Of course, at this stage the prospective contractor has not submitted any detailed information regarding how he arrived at his tender sum – only the lump sum figure has been stated. Thus, prior to commencing the project, and assuming that no detailed priced 'bill of quantities' has been used, it would be reasonable to ask the successful contractor to submit a detailed breakdown of how he arrived at the quoted price. This is especially important on smaller domestic projects, where very little detailed information regarding quantities has perhaps been stated in the tender documents given to the contractor. Its usefulness becomes potentially more important later in the project (see section 6.3.5 below – 'Provision for variations').

There is often speculation on the number of companies that should be invited to tender for a project and, clearly, there is no single correct answer to this question. Ideally, common sense should rule – generally the smaller the project, the fewer the number of tenders that should be sought – but for most small- to medium-sized projects, between three and six tenders should be satisfactory for most instances.

Many domestic clients wishing to construct, say, a house extension will not, however, follow the *formal* tender process described above. Instead, they will in all probability simply invite a small number of local builders to provide quotations for their project in the form of a quotation letter – the only other contract documents exchanged being the project drawings and specification (usually the same drawing that was used to obtain Building Regulations approval). This is satisfactory as a general principle, as a written quotation on the contractor's letterhead would commonly be quite adequate as the basis of a legal contract between the parties.

However, care should be taken to ensure that all quotations received are in relation to the provisions of the same level of service and materials. For example, will the contractor within that price do other works ancillary to the construction of the new extension but that are not stated on the drawing or in the specification? A recent court case for a domestic extension project revolved around the fact that the client refused to pay a portion of the agreed quotation because the contractor had not laid new paths to the perimeter of the extension. Such works were not shown on the project drawings and the contractor's quotation letter stated clearly that the price was for the works ' *as shown on the project drawing*'. The client alleged to the court that during his initial meeting with the contractor he had asked that a path be included in the quote. Whether the contractor deliberately ignored this request or not is unknown but, nevertheless, the contractor denied in court that any such request had been made. The court found in favour of the contractor on this occasion and the client was left to rue the fact that he did not include the requirement for this additional work in the tender documents.

Another potential pitfall for the unwary or inexperienced client concerns the use of the phrase 'estimated cost' within any quotation. Occasionally, there are elements of any project that are difficult to accurately quantify at the outset for one reason of another. Obvious examples might be for the supply and installation of a single large fitting – a bathroom suite for example. Many householders will not have decided at that time precisely which make/model of suite they would like fitted and so the contractor preparing a tender has little option but to incorporate a lump sum figure to represent the likely amount that this element would cost. Such an inclusion is called a 'prime cost sum' and, ideally, all of the contractors submitting tenders for the works should be instructed in the tender documents to include the same specified amount – say, £2,000 or whatever is deemed to be appropriate. This helps to ensure that all the tenders are able to be compared as being 'on a level playing field' as it were.

Another means of achieving the same goal relates to items where the precise quantity of work is unknown at the outset – an example might be to allow for the cost of exposing a buried drain pipe for inspection, to determine whether it needs protecting from the proposed works. In this instance, a 'provisional sum' should be specified (again, ideally by the client at the outset). However, in such cases, the work content that is represented by the 'provisional sum' is usually only able to be undertaken with the express written permission of the client or the client's representative. Such works would then probably be done, say, on a 'dayworks' basis, where the contractor is reimbursed for the actual hours the work takes plus the cost of materials. The hourly rate for such works would either be agreed as part of the tender or, alternatively, would be established by reference to nationally agreed hourly rates for the relevant trades involved.

Thus, other than for the typical instances discussed in the preceding paragraphs, any 'quotation' letter that includes the phrase 'estimated cost' should be either clarified, or viewed with a degree of caution. A case is referred to elsewhere in this book, where a domestic client appointed a contractor to build a large extension to his home, plus some other ancillary external landscaping works. The matter ended up in court because the contractor submitted invoices for some 35 per cent more than the original 'quotation' to which the client considered he had agreed. The case was fairly complex for reasons that are irrelevant in the context of this section; however, the relevant point to be made here is that the letter commenced with the heading (in bold capital font) that the document was a 'QUOTATION' for the erection of the extension at the named property, but finished at the bottom of the second page with the statement 'Estimated Cost', followed by the tender amount. The two parties' respective solicitors ultimately spent weeks arguing over whether the contractor's price was truly a 'quotation' or an 'estimate'. One can only imagine the legal fees that the case finally attracted and, as so often happens in such matters, neither party ultimately escaped with a healthy bank balance.

6.3.5 *Provision for variations*

There is a maxim that it is much less expensive to spend time and money at the design stage, ensuring that a project is as near to 'perfect' (in the client's eyes) as possible, than after the works have commenced on site, because it will be much more expensive to change details of the design. There are a number of reasons why this might be true, including for example:

- There may be costs incurred in removing elements already installed, which are now to be replaced with an alternative product or detail.
- There may be related design fee costs incurred for amending the project drawings/documents to suit the new product or detail.
- There may be related fee, etc., costs if the amended design requires submission for, say, local authority approval (e.g. under Planning or Building Regulations legislation).
- There may be related delays in completing the project, resulting in additional contractors costs for general site preliminaries, etc.

Also, when initially tendering for a project, the potential contractor is in competition (in most instances) with other companies and will be likely to be keen to win the order by submitting the most cost-effective tender (regrettably, this is too often simply interpreted by the client as the 'cheapest'). However, once the contract has been awarded, the successful contractor is under much less incentive to be quite so keen when calculating the cost of any variations to the original design.

On larger projects, or where the tender has been based upon an itemised 'bill of quantities and specification', there is much less opportunity for an unscrupulous contractor to take advantage in this way. Unfortunately, however, as stated in the previous section, for the majority of smaller projects (particularly those for domestic customers) no itemised bill of quantities is used – the prospective contractor instead basing his quotation purely upon the project drawing (usually the same drawing that was used to obtain Building Regulations approval) and its attendant specification for the works.

As with the previous section of this chapter ('The tender process', above), for any private householder who is intending to embark upon building, for example, a house extension, there is a 'golden rule' to be observed when negotiating variations to the content or specification of the works. Put very simply, it is to remember to *confirm any changes in writing* to the contractor. A professional project manager or supervisor acting on behalf of a client under the provisions of the JCT MW form of contract would not even contemplate agreeing any changes to the specification or content of the client's project with the contractor without confirming the amendments by issue of a *written variation order*. This is simply a document setting out (in an appropriate amount of detail) what the changes to the works comprise and, most importantly, confirming what the financial effect upon the total contract price will be, if any.

Far too many domestic construction projects are undertaken with no exchange of correspondence following the initial quotation, and in far too many cases this results in an unnecessary argument – because three months after the alteration to the specification was verbally agreed upon between the client and contractor, one of the two parties will remember the discussion slightly differently to the other. This is a potential argument that can be avoided by the simple expedient of one party (usually the client) giving a simple written note to the other confirming the details of the changes and the agreed price variation.

This is the point at which, the need for the contractor to have provided at least some form of cost breakdown is important. If the contractor has provided a detailed analysis of his tender build-up down to each individual work item, it is a relatively simple process to compare this document with the contractor's subsequent quotation for the cost of any variations. Even if the contractor has only provided a very simple breakdown of costs into principal work elements (e.g. ground works, brickwork, roof structure, internal joinery, plumbing, electrics, plastering, etc.) this will serve to restrict the scope for the contractor to subsequently be too creative when pricing-up any proposed variations – though obviously the fully detailed tender estimating sheets would be much better for this purpose, if the builder has them.

This is not, as they say, 'rocket science' – if a contractor has included in his original quotation for a prescribed number of fittings for a consistent unit price (say, electrical sockets at £30.00 apiece) then it follows that any additional fittings of that type should be expected to be priced on a similar basis. If the variation quotation comes in at a higher unit price that in the original tender quotation, the client is able to ask the contractor for an explanation of the difference in unit price.

6.3.6 *Payments*

The basis for reimbursement of the contractor as the work proceeds should be clearly set out in the tender documents. Dependant upon the project, that could be on the basis of a single payment upon completion or, more likely, on the basis of a series of phase payments – either upon completion of specified sections of the works or, more commonly, at regular time intervals throughout the life of the project, in which case the amount due would be subject to an agreed assessment of the percentage of work completed at that time, for each major element of the project.

The JCT MW form of contract sets out at Clause 4.3, the basis for interim payments of this kind, as follows:

> 4.3: The Architect/Contract Administrator shall, at intervals of 4 weeks calculated from the date of commencement of the works, certify progress payments of the percentage stated in the contract Particulars of the total value of:
>
>> 4.3.1: work properly executed, including any amounts ascertained or agreed under clauses 3.6 and 3.7 (i.e. variations and provisional sums – see earlier sections) and;
>>
>> 4.3.2: materials and goods which have reasonably and properly been brought onto the site for the purpose of the works and are adequately protected against weather and other casualties. . . .

It would also be common practice to withhold a percentage of the monies due to the contractor in the form of a 'retention' against the possibility that any significant defects might become apparent during the course of the remainder of the project works and for a fixed period after the project works have been completed. This 'rectification period' should, again, be clearly set out in the tender documents and for the majority of small- to medium-sized projects, the rectification period will be from three to six months. For certain larger or more complex projects, it may be appropriate to state a longer retention period than this, but it is unusual for the rectification period to exceed 12 months.

The JCT MW form of contract sets out the formal basis for this retention of monies from the contractor, which, unless stated to the contrary at the tender stage, is taken by default under that form of contract to be 5 per cent of all interim payments and 2.5 per cent of the final payment at the completion of the works (i.e. at the commencement of the rectification period).

On domestic projects, very small building firms will commonly request that the client make them a payment in advance, under the premise that this money is to be used 'to buy materials'. No doubt there are some very genuine, trustworthy building companies in the marketplace who use the money for just that purpose; however, from the number of reports one hears every year, there are too many contractors who falsely obtain this money in order to pay their operatives to complete the job they are still working on for someone else – who has not yet paid them.

Most established building companies will have an account with their local builder's merchant, giving them up to 30 days (or more) credit for any materials delivered, and one wonders, therefore, why they should need to have significant sums from the client for the purpose of 'buying materials'? For the domestic client, there are two simple steps one can consider when the client's preferred builder still demands a payment in advance for materials before commencing work: first, offer to make a payment to the contractor immediately upon delivery to site of the materials for the project. The amount of that payment would be determined by the delivery note or invoice from the builder's merchant to the contractor. If such a method pf payment is still not acceptable to the contractor (perhaps he is unable to fund the short delay in payment with the merchant) then offer to visit the builder's merchant with the contractor in order that you may pay the builder's merchant directly for the materials – perhaps using your credit or debit card. The important factor here is that, once you have paid for the materials and they are delivered to your property or site, they are your property and in the unthinkable event that anyone (including the contractor) should seek to remove some or all of those materials from your property for any reason that would constitute a theft.

6.3.7 Disputes

Where no form of contract such as the JCT MW has been adopted, the parties are likely to have to rely upon common law for ultimate resolution of any disputes and this is one of the primary reasons why it is preferable to use one of the recognised forms of contract for any project, no matter how small.

The JCT MW form of contract makes clear provision for those instances where the parties are unable to agree on a particular item. Section 7 of the form of contract provides three alternative means of pursuing such a dispute, either via mediation, adjudication or arbitration.

- Mediation, sometimes referred to as ADR ('alternative dispute resolution'), is a process whereby an independent third party (the mediator) is engaged to try to help the parties to find some 'common ground' in order to resolve their dispute. By its nature, mediation can only ever be successful if both parties come to the proceedings with a real intent to compromise to some degree. It differs from other types of dispute resolution, which tend to concentrate on deciding 'legal rights', whereas mediation has perhaps been best described as 'trying to find a solution that the parties can live with'.

- Adjudication is a relatively recently introduced form of dispute resolution, whereby either party may submit a matter for adjudication by an independent third-party adjudicator. It was introduced by the provisions of the Housing Grants Construction and Regeneration Act 1996 and provides a statutory right to access this dispute resolution process which, it is intended, should be completed within 28 days – within which timeframe the appointed adjudicator has to decide upon the rights of the parties under the building contract. The adjudicator's decision is binding upon the parties and can only be overturned, revised or confirmed in arbitration (see below) or in litigation. The primary benefit to the contractor in calling for adjudication is that the matter becomes settled quickly, thus potentially restricting the effect upon cash-flow. However, the principle behind adjudication is that each party meets their own costs in the matter, irrespective of whether they 'win' the decision or not.

- Arbitration is not as some may believe a recent innovation, but in fact pre-dates modern legal systems and courts, finding its roots in Roman times. The current process is founded in the Arbitration Act 1996 and provides a less rigid and hopefully less costly alternative to the procedures that existed before. In the JCT MW form of contract, the procedures for arbitration are set out at Schedule 1, Clauses 1 to 6. The arbitrator's decision is final and binding upon the parties, except that the parties may apply to the courts to determine any question of law arising in the course of the reference to arbitration and they may appeal to the courts on any question of law arising out of the arbitrator's award.

6.4 Responsibilities

It is fundamental to the success of any relationship that all the parties understand and live up to their individual and collective responsibilities. This section of this chapter of the book will examine the relationships and responsibilities of the principal parties to the construction contract.

6.4.1 Client

This may seem just too fundamental a point to make, but surely the primary responsibility of the client is to ensure that adequate finance is in place to fund the cost of the whole project prior to commencing work. The phrase 'adequate finance' is intended to include *all* anticipated costs, plus a provision against any unexpected costs – a 'contingency fund'. The level of 'contingency fund' that one should provide will of course vary with each individual project and may be as little as 5 percent up to, say, 25 percent in some cases – but a more common figure is 10 percent of the tender sum. Please note, this is not a contractual obligation, but plain common sense.

During the course of any project, something is potentially likely to go wrong, no matter how careful and conscientious the contractor and his operatives try to be. It is invariably the fact that contractors employ human beings to carry out the works on site and thus there is always a capacity for error or mishap. The important thing for the client to do here is not to immediately take up a contrary position to the contractor, but to work with the contractor to find a solution to the problem.

Too many clients – and, regrettably, this comment applies to most domestic clients – expect the contractor to be able to achieve 'perfection' in every aspect of the finished works.

Of course, 'perfection' is something only rarely attained in any walk of life. This is not to say that one should accept 'second best', but one has to understand that the basis of most UK law is that of 'reasonableness', and a 'reasonable' standard of workmanship is not necessarily consistent with 'total perfection'. Reference was made earlier in this chapter (see section 6.1.2 above) to the current British Standard to BS5606:1990, 'Guide to accuracy in building', which clearly states that compliance with that document does not require perfection to be achieved in plan dimensions, verticality, straightness or level.

It has been said many times that the British Standard specifications are *minimum* acceptable standards and not necessarily representative of *best practice* in all respects – and this is, of course, a fair comment. So, a client who wishes to achieve a higher standard of finish or accuracy in construction must specify these requirements in the tender documents. However, beware the client who demands unreasonable (i.e. practically unattainable) levels of accuracy or finish, because it is extremely unlikely that any subsequent formal dispute process would support the client's 'right to perfection' unless such were apparently achievable in every case.

The client also has more formal responsibilities under the Construction (Design and Management) Regulations 2007 – these are discussed in more detail in section 6.6 of this chapter, below.

6.4.2 *Contractor*

At face value, the contractor's responsibilities would seem to be fairly straightforward – they involve completing the project to the client's reasonable satisfaction, making sure that all necessary insurance provisions, etc., are in place before commencing. However, there is a fundamental question that is often neglected – is the contractor suitably experienced, qualified and with sufficient financial resources to undertake the works?

Just as the client has a responsibility to ensure that sufficient funding is in place to pay the contractor, so the contractor must be able to 'cash-flow' the project for the interim period between construction of an element on site and receipt of payment for that element from the client. The JCT MW form of contract states at Clause 4.3 that 'the final date for payment of the amount certified shall be 14 days from the date of issue of the certificate'. However, any interim certificate could include works that were undertaken, say, four weeks previously (i.e. immediately after the previous interim certificate was assessed) and so such payment might be some six weeks or even more after the contractor incurred the cost of the labour and materials comprising that work item.

'Horses for courses' is an old adage that is applicable here. Just as there are professionals designers who specialise in, say, historic buildings, modern multi-storey buildings, etc., there are contractors who have specialist expertise in certain types of project. It is unlikely, perhaps, that a small firm of builders would necessarily have the expertise to undertake the erection of a steel portal framed industrial building and, indeed, they may not have sufficient depth of financial acumen to be able to fund the project. Similarly, a large contracting firm with the substantial overheads associated with a successful organisation manufacturing and erecting high-quality pre-cast concrete building products would be unlikely to be able to cost-effectively and profitably undertake a kitchen extension for a domestic customer. These are, of course, deliberately extreme comparisons just for the sake of making a point; however, the principles apply for any project. The client or client's representative should always ask the contractor for references from past clients for this type of project, but the contractor also has an overriding responsibility – one might even describe it as a professional duty of care – not to submit a tender for a project that he/she is aware is beyond the company's capabilities to satisfactorily complete.

6.4.3 *Designer*

The extent of the designer's responsibilities may vary, depending upon the terms of the appointment by the client, and so the following comments are intended to give an overview of the designer's potential responsibilities, some or all of which he may be encumbered with on any given project.

Principally, there is a responsibility upon the designer to provide the client with a project design that not only matches the design 'brief' for aesthetic appearance and practical performance, but that also complies with all relevant legislation and, most importantly, that is able to be built. By 'able to be built' one infers not only that the project should be physically able to be constructed, but that it is able to be constructed within the limits of the client's available financial resources.

The designer's responsibilities might also extend to identifying suitable contractors for the tender list; preparation of tender documents; administration of the tender process, including an appraisal of the returned tenders; and advising the client on final choice of contractor. During the design phase, the designer may be required to provide a degree of liaison with other professionals, including perhaps architectural designers, landscape architects, structural engineers, mechanical/electrical engineers, quantity surveyors, party wall surveyors, local authority planning officers and building control officers, etc.

Once the project is under way, the designer might also provide the function of site supervision and quality control of the works – or alternatively, may assist the client in identifying another suitable firm or perhaps appoint a clerk of works to do this.

The designer also has statutory responsibilities under the Construction (Design and Management) Regulations 2007 and these are discussed in more detail at section 6.6 below.

6.5 Special conditions

Every project is different to one degree or another, giving rise to certain difficulties that have to be overcome by the designer, the client and/or the contractor. The difficulties are sometimes related to the site – its size, geographical location, topography/terrain, proximity to other properties – or perhaps there are pseudo-political issues pertaining to the proposed usage of the site. Sometimes the difficulties might be simply due to the complexity of the project, especially if the work is considered to be 'groundbreaking' (no pun intended) with regard to some fundamental aspect of the manner in which the project is to be designed, funded or constructed.

Whichever of the above, and possibly other factors, may be relevant, any special conditions should be identified and measures put in place to accommodate them as early on as possible in the project and certainly prior to commencement of the works on site if at all possible.

The majority of small- to medium-sized projects are likely to have only one or two elements that could be classified as 'special conditions' in the context of this section. Most of these have been addressed separately in Chapter 4 of this book and thus there is no need to repeat them here. However, it is important that any outstanding issues that have not been addressed by the designer should be brought to the attention of the prospective contractor at the tender stage, making it clear that the responsibility for resolving these factors during the course of the construction phase will fall to the contractor.

6.6 CDM Regulations

This section will examine the principles and objectives of the Construction (Design and Management) Regulations 2007 (CDM 2007) and the responsibilities that they impose upon the client and designer. It is not intended that this section will provide an in-depth review of CDM 2007, as there are sufficient other publications available for that purpose. Rather, it is intended to be a brief overview of the main factors that will affect the majority of small- to medium-sized construction projects.

6.6.1 *The objectives of CDM 2007*

CDM 2007 replaced both the former Construction (Design and Management) Regulations 1994 and the Construction (Health, Safety and Welfare) Regulations 1996, combining them into a single parcel of legislation. The principal objective of CDM 2007 and the supporting document, the Approved Code of Practice (ACoP), is to reduce the incidence of construction accidents and ill-health. They are intended to focus attention on the planning and management of construction projects from design concept onwards. The essential aim is for health and safety considerations to be treated as an essential, intrinsic part of each project's development and not just an afterthought.

The main objectives of CDM 2007 are:

- To simplify the regulations and improve clarity – to make is easier for duty holders to know what is expected of them.
- To maximise their flexibility – and enable the regulations to be applied across a broad spectrum of different contractual arrangements.
- To encourage individuals to focus on planning and management – i.e. not just upon the paperwork, but to emphasise the use of active management and effective risk control and discourage the use of endless 'paper chases' and unread plans.
- To strengthen the requirements on co-operation and co-ordination, thus encouraging better integration, particularly between designers and contractors, so that they can share problems and find solutions to the problems, *before* they materialise on site.

The key aims of CDM 2007 are:

- To have the right people for the right job at the right time – to manage risks on site, reduce paperwork and encourage teamwork.
- To ensure people appointed are competent to do the work.
- To reduce bureaucracy and paperwork.
- To focus on the effective planning and management of risk.

CDM 2007 requires that the details of certain projects are formally notified to the Health and Safety Executive (HSE). Except where the work is for a domestic client, the HSE must be notified of projects where construction work is expected to:

a) Last more than 30 working days; or
b) Involve more than 500 person days (for example, 50 persons working for over 10 days).

The Approved Code of Practice (ACoP) to the Regulations sets out a summary of the duties of the principal parties to each construction project and these are discussed in the successive sections. However, it is the duty of all parties to a construction project, irrespective of whether the individual project is notifiable to the HSE or not, to:

- Check their own competence to undertake the task.
- Co-operate with others and co-ordinate work so as to ensure the health and safety of construction workers and others who may be affected by the work.
- Report obvious risks.
- Comply with requirements in Schedule 3 and Part 4 of the regulations for any work under their control.
- Take account of and apply the general principles of prevention when carrying out duties.

6.6.2 Client's duties under CDM 2007

Unless the project is for a domestic client, then for all other construction projects, CDM 2007 requires that the client shall:

- Check the competence and resources of all persons/organisations appointed under the regulations.
- Ensure that there are suitable management arrangements for the project, including welfare facilities.
- Allow sufficient time and resources for all stages of the project.
- Provide pre-construction information to designers and contractors.

Where a project is deemed to be 'notifiable' (i.e. where the HSE must be formally informed about the project), the client must, in addition to the duties above, also:

- Appoint a CDM co-ordinator (see section 8.6.4 below).
- Appoint a principal contractor.
- Make sure that the construction phase does not start unless there are suitable welfare facilities and a construction phase plan is in place.
- Provide information relating to the health and safety file to the CDM co-ordinator.
- Retain and provide access to the health and safety file.

A client is deemed by the regulations to be 'an organisation or individual for whom a construction project is carried out'. This can include, for example, local authorities, school governors, insurance companies and project originators on 'private finance initiative' (PFI) projects. Note that, as stated above, domestic clients are a special case and do not have duties under CDM 2007.

6.6.3 Designer's duties

Unless the project is for a domestic client, then for all other construction projects CDM 2007 requires that the designer shall:

- Eliminate hazards and reduce risks during design.
- Provide information about any remaining risks.

Where a project is deemed to be 'notifiable', the designer must, in addition to the duties above, also:

- Check that the client is aware of his duties and that a CDM co-ordinator has been appointed.
- Provide any information needed for the health and safety file.

A designer is deemed by the regulations to be

those who have a trade or business which involves them in:

(a) Preparing designs for construction work, including variations. This includes preparing drawings, design details, specifications, bills of quantities and the specification (or prohibition) of articles and substances, as well as all the related analysis, calculations and preparatory work; or,

(b) Arranging for their employees or other people under their control to prepare designs relating to a structure or part of a structure

It does not matter whether the design is recorded (for example on paper or a computer) or not (for example it is only communicated orally).

6.6.4 CDM co-ordinator's duties

If the project is for a domestic client or if the project is not one that is notifiable to the HSE, then it is not a necessity for the client to appoint a CDM co-ordinator. Where a project is notifiable, the CDM co-ordinator shall:

- Advise and assist the client with his/her duties.
- Notify the HSE.
- Co-ordinate health and safety aspects of design work and co-operate with others involved with the project.
- Facilitate good communication between client, designers and contractors.
- Liaise with the principal contractor regarding ongoing design.
- Identify, collect and pass on pre-construction information.
- Prepare/update the project health and safety file.

7 Specification clauses, examples, sources and how to draft

Risk in construction is a constant menace. Unknown features come to light as work proceeds, affecting both cost and time. The allocation of risk can become one of the main areas of dispute between the parties (see Chapter 8).

We have already had a brief journey through the legal provisions of tendering and obtaining quotations and the way in which statute moulds and shapes the design, but what about the contractual and works content? This is written and codified in a document called a 'specification'. One of the main purposes of a specification is to *minimise risk* by a pre-contract discussion of sharing that risk between the parties and the specifier.

As will have been seen from the previous chapter, preparing adequate documentation is an essential preparatory stage to getting building works executed. The aim here is to minimise risk.

The expectations of a client or a contractor are likely to be wholly different at the beginning of a project when details have not been articulated or explored. As an example we can use a client's desire to refit and alter a bathroom. The bald statement 'I want a new bathroom' has to be discussed and developed and then codified to ensure that all parties hold the same mental picture. It is only then that the specification for that picture can be developed.

7.1 Finding and using resources

a) British Standards

These are produced by the British Standards Institution, the UK's standards organisation, and often include or are supplemented by Europe-wide or even worldwide standards. These set out recommended standards for supply and manufacture of materials and components. It is not necessary to buy all the ones covering the project works, but if particular ones are referred to in the specification, they should be available to all the tenderers for inspection. Often, the requirements of the Standards are well known in the industry and copies are not required.

b) Codes of practice

These are also produced by the British Standards Institution and often include or are supplemented by Europe-wide or even worldwide standards. These set out recommended standards of workmanship and methods to be followed when using materials. As with the British Standards, it is not necessary to buy all those covering the project works, but if they are referred to in the specification, they should be available to all the tenderers.

c) Approved documents
    As noted earlier, these are published by the UK Government and set out means of achieving compliance with Building Control standards. These are not obligatory and it is possible to achieve compliance without using these suggested solutions. They can be obtained from the appropriate department (see: http://www.planningportal.gov.uk/uploads).

d) Manufacturers' documents

    i) Catalogue
        These are often 'glossy' publications and do not always provide methods of application or details of the components.

    ii) Instructions
        Useful for the specifier to see how the material or component is to be used, e.g. setting or drying times, details of which can then be incorporated into the specification.

    iii) Limitations
        A degree of cynicism is required to tease out the implications for the client and contractor/tenderer of the use of particular components or materials. One that looks wonderful may be awful to use.

    iv) The British Board of Agreement
        The BBA is the UK's major authority offering approval and certification services to manufacturers and installers supplying the construction industry. Certificate test results are published, setting out standards and applications for materials and components.

    v) Internet research.

    This is always a swift and useful source of up-to-date information.

7.2 The project works

The bald statement 'I want a new bathroom' does not convey the detail required to achieve the desired result. The bathroom may be a simple box room equipped with a plain bath and basin. On the other hand it may be a marble-tiled room with luxuriant fittings and enough gadgets to make the user feel like they are entering a car wash.

In order to obtain accurate and comparable quotations for the work, all parties must settle on a single set of works so that comparison between the various tenders received can be made on the same basis.

Standard documentation is available from several sources. In the UK the National Building Specification, text books, manufacturers' documentation and other proprietary literature all feed ingredients into the mix to produce a specification.

Taking any of these ingredients in isolation and without a critical eye is fraught with difficulties. It pays dividends in the long run to adopt a cynical attitude towards each source. Manufacturers seek, naturally, to specify their own products and to link their solutions to other compatible products. So, with this somewhat cynical viewpoint in mind, a route through the 'dark mystery' is required. To do this, ensure that notwithstanding the actual aims of the parties, they are divided into their different roles, e.g. client, end-user, contractor, etc., and of course the specifier. So, for this to happen, there must be a 'specifier' or a manager taking responsibility for, and holding the overall vision of, the project.

Having identified the parties, the role of each must be codified and their roles confirmed, if possible, by agreement, or if not, by being imposed by the specifier.

'The project' is the title given to the whole scheme, both building works and human interaction. This book is primarily concerned with the former, but inevitably the latter cannot be ignored.

7.3 Deciding on the 'specifier'

It is extremely difficult to separate the experiences and aspirations of the author from what is written. The separation of roles allows for the posing of critical questions to those either seeking works or seeking to influence the works specified.

The 'specifier' requires technical knowledge but perhaps more importantly an understanding of materials and how components fit and work together. For example, he or she should understand how metals interact and the inappropriateness of mixing iron and copper, which gives rise to electrolytic action, a basic circumstance to avoid.

As professionals, the authors are naturally inclined to use professional advisers, but that is not always necessary. What is essential is that there is a 'referee' to assemble and filter conflicting demands and desires. The 'referee' will need to have technical knowledge in order to fully evaluate the varying demands. For this reason the 'specifier' should be suitably informed/qualified person

There will be areas of works, for example services installations, which require specialist knowledge or where particular products cannot be readily evaluated without research. In an ideal world, the specifier will be able to assess the diverging claims of various solutions, but that may not always be possible

Before specifying alterations it is essential that the need for repair is considered.

7.4 Assessing repairs and priorities

Buildings are living organisms, part machine and part organic. We would like to think that they are wholly mechanical, inert and free from unpredictability. They are not. In order to have an appreciation of how they are affected by the use to which they are put, by external influences such as the weather and by their invited and uninvited occupants, it is first necessary to appreciate what they are made of.

If you do not know what a house is made of, you will not understand either the nature or significance of a defect, or how a particular part of the building contributes to the overall performance of the dwelling as a shelter.

It is suggested that one of the marks of modern society and perhaps of civilisation is that we have sophisticated organisations and solutions to common human problems. We have evolved a system of construction intended to provide healthy buildings within which we work and live, shielded from the elements. This basic need to be protected from natural risks and phenomenon has been recognised as a primary demand with the provision of a home being elevated almost to a basic human right, alongside liberty and freedom of expression.

In looking at buildings and their defects it is always worth bearing in mind the essential purpose for which they were created. The question of whether a building can perform the function for which it was intended is always a relevant consideration.

A building is like an onion, with one layer covering another. In this chapter we will examine what materials are used, how things are put together and what the enemies of those materials and elements can do.

7.5 The materials of construction

Materials used for buildings may often appear to be inert. However, we have yet to develop materials for common use which are not adversely affected by the earth's natural environment. In this section we look first at the nature of the primary materials from which we continue to build houses.

7.5.1 *Timber*

Probably the most commonly found material in residential buildings, this has been the basic material for construction since humans relinquished the natural shelter of caves.

Timber comes from trees. It is classified under two headings: hardwood and softwood. These groups are named not for the inherent feel of the timber, although in some cases that follows the name, but because of the structure of the wood. Pine is a softwood, quite hard and durable, but Balsa, malleable, soft and compressible, is a hardwood.

Hardwoods take longer to grow and generally are not capable of replacement at anything like the rate they are used. Softwoods tend to be fast-growing and renewable. With the realisation that the earth's resources are running out, builders have steered away from hardwoods except for particular uses where no alternative is suitable.

Timber is seasoned, i.e. dried, to a low moisture content prior to use. The lowering of the moisture content reduces the risk of warping and shrinkage once it is installed in a dwelling and the heating is turned on. It also reduces the moisture content to a level at which it is insufficient to promote or sustain fungal growth.

In domestic buildings, softwoods pre-dominate. The nature and quality of the timber will vary. Older buildings, particularly in the mid-Victorian period, used well-seasoned softwoods, durable and resistant to fungal and insect attack (see below). As building production increased in the later Victorian era and at various times this century, the demand for good, well-seasoned timber has out-stripped supply. To satisfy demand, younger or less durable softwoods have been used.

After years of failure at the hands of fungus and insects, the building industry accepted that poorer quality timber could only be used if it were treated with chemicals to resist more effectively the advances of those wishing to devour it. Today, almost all softwoods used in buildings are pre-treated with chemicals to hinder the ravages of fungus and insects.

Timber is used in two qualities in construction. The cruder, un-planed timber is used for the structure – the floors, the roofs and the parts which will not be seen when completed. This does not permit the builder to use lower grade or strength of timber, merely that which is not smoothed to an aesthetically acceptable finish. Planed timber, shaved to a smooth surface ready for staining or painting, is used for the visible joinery in the building, ranging from floorboards to furniture.

The excess of demand over supply, with consequent increases in timber prices, also lead to the search for new materials using otherwise rejected chippings, off-cuts, etc. Composite, factory-made timber materials are now commonly found in buildings. The most common is probably particle board (the generic term for chip-board, laminboard and a myriad of others). These boards are manufactured with off-cuts, shavings and otherwise rejected pieces of sound timber, mixed and bonded together with resin to form a dense and solid mass. The density of the boards varies, from, highly compressed hardboard to soft insulation board. These boards are used for flooring, kitchen units, shelving, doors and mouldings. Although apparently different to natural timber, they have the same properties and vulnerabilities to fungus and insects.

7.5.2 Brick

Do you remember the Israelites making bricks with mud and straw as Charlton Heston and Yul Brynner looked on in *The Ten Commandments* – probably too long ago! This method of making bricks is no longer used in the UK, but throughout the southern hemisphere, these sun-dried (adobe) bricks are the basic building material.

In the UK we have large deposits of clay. Clay, when baked, forms a hard, dense and stable material which can resist the effects of rain and temperature and humidity changes. Since Roman times, bricks for the better quality and eventually all buildings have been of baked clay. Modern bricks are manufactured to regular sizes and achieve a remarkable degree of standardisation of quality, thanks primarily to the improvement on oven-baking techniques. Clay bricks are at their driest when they leave the kiln. From that point they will absorb moisture, although the extent of absorption will diminish with time. As the moisture content varies so will their size, causing movement within the structure.

In more recent years, other materials such as calcium silicate have been used for bricks. Unlike clay bricks, the movement of these is primarily generated by drying rather than wetting. Their use in the ground therefore presents no problems, but above ground severe movement can occur.

In the nineteenth century it was still necessary to have a skilled assessor of bricks to select those which were of better quality and to allocate those bricks that were inferior to a suitable task. Today, this selection is done not from each batch but from the catalogue.

Because of the regularity of size and shape, bricklaying has become cheaper and faster, and today, notwithstanding the increasing use of other structural materials, brick is still holding its own as a major constituent of domestic buildings.

Bricks vary in density and strength. Some are extremely dense and almost impervious. These can be used as a damp-proof course or for below-ground work where dampness cannot be avoided. Above ground, less dense bricks with varying degrees of porosity. are used.

Because high density, impervious bricks are expensive to produce, lower density bricks are far more common. Buildings are designed to accommodate this and it will be recalled that cavity walls assume the possibility of water penetrating the outer skin of brickwork.

In most cities, older buildings are built of 9" brickwork (i.e. one brick thick). If these walls were in an exposed position, or if there was a defect such as a leaking rainwater pipe allowing water to saturate the exterior, then water penetration to the interior would be likely. Thicker solid walls, denser bricks and cavity construction are all factors which will improve the wall's resistance to such water penetration.

7.5.3 Mortar and pointing

When built as a wall, bricks are bedded in mortar. This is a composition of cement, lime and sand in varying proportions. The mix, with water, is designed to hold the bricks together yet to allow some flexibility. More modern mortars mixes are often hard and whereas with older buildings, some structural movement can be accommodated, in modern structures, every slight distortion can be marked by a crack.

When pointing is missing, soft or perished, water can more readily enter the top surface of the brick. Combined with frost action, this will lead to breakage of the brick and increase the possibility of water penetrating to the interior.

7.5.4 Render

Cement mixes are often applied to the outside of buildings. In some cases this is for aesthetic reasons only, but more usually it is to provide a weather-proofing function. The mix will, like mortar, be of cement, lime and sand and should provide a flexible surface. Again in modern times, hard inflexible render mixes have been used. Crazing of render with minor, non-structural cracks will occur with less flexible mixes or where the render is stronger than the brickwork on which it is applied. This crazing will allow water to penetrate and coupled with frost action will cause deterioration of the brickwork often before the render itself manifests any serious failure.

7.5.5 Insulation

Modern buildings are required by the Building Regulations to achieve a minimum insulation value. For cavity walls, this is usually achieved by the insertion of insulation in the cavity. The materials commonly used are rock fibre (fibre glass), polyurethane foam and urea-formaldehyde foam.

Urea-formaldehyde foam has achieved some notoriety for discharging gases into the dwellings. The chemical reaction of the two agents which make up the foam releases formaldehyde gas. The cavity must be well ventilated at its head so that this gas escapes.

Other risks with insulation in cavity walls arise from bridging of the cavity, allowing rainwater to pass from the exterior to the interior (see below).

With solid walls, internal or external insulation is applied over the brick. The increased thermal efficiency will only be risk free if any such insulation takes full account of moisture passing through the wall.

7.5.6 Rock

Rocks, either cut or crushed supply a main source of building material. When used as a prime material without chemical conversion it is commonly termed stone (see below)

7.5.7 Concrete

The chemical conversion of crushed stone into cement was perfected in the last century. Today there are a multitude of different cements which are used to make concrete. In all cases the cement is mixed with sand and stones (together called aggregate) and water. The mixing is both a physical blend and a chemical change which causes the cement to heat and cure into a hard and durable material.

In most domestic buildings ordinary Portland cement is the main constituent of concrete. It develops its strength over a week or two. It has little resistance to acids or sulphates and may not always be used below ground where sulphate resistant concrete are required. Other cement mixes achieve more rapid hardening or are especially resistant to acid or sulphate attack.

High alumina cement was rarely used prior to the 1950s. Its attraction was that it developed very high strength within 24 hours, thus speeding up the building process. It was particularly used in factory manufacture of pre-fabricated beams. After placing, high alumina cement concrete undergoes a chemical change. This is known as 'conversion'. The conversion will result in a reduction in strength over 5 or 10 years. When converted the concrete is vulnerable to acid, sulphate and alkali attack. After 10 years some slight increase in strength occurs, but in warm, moist situations, further chemical action is possible depending on the aggregates used.

7.5.8 *Plaster*

The principal reason for applying plaster to wall surfaces is to conceal the unavoidable irregular surfaces of the structure itself. Plastering achieves this by filling the gaps and levelling out the surface. It also adds to the thermal efficiency of the wall or ceiling and to the sound reduction.

Where a wall has been damp, for example when affected by rising damp, specialist plasters are applied to resist the passage of residual salts to the interior, which would damage the decorations.

7.5.9 *Stone*

Stone is the term commonly applied to rock which is not chemically converted or otherwise restructured. It is rock which is cut from the natural source and split and shaped (dressed) and then used in its natural state as a building material.

Stone has been used for thousands of years as a building material. Although thought of as inert, stone may contain organic material and varies from very dense and strong (e.g. granite) to soft and porous (e.g. sandstone). The durability will vary significantly from one quarry site to another.

Slate is also a stone and has been used for many years as a roofing material and as a damp-proofing medium. Due to its high cost and finite resources, imitation slates have been manufactured. Until the mid-1970s these were primarily of asbestos-cement but today are of cement and mineral fibres.

Deterioration is generally slow and is aggravated by air pollution. Natural slate should have a life expectancy of between 50 and 80 years and imitation slate perhaps 40 to 70 years. It is more likely that deterioration will be caused by breakage of the slates through impact or corrosion of the metal fixings. With old slate roofs, iron nails fixed through the slates will have corroded well within the lifespan of the slate, causing the slates to slip. This is sometimes known as nail sickness. More modern roofs should have fixings of non-ferrous metals which will be less affected by corrosion.

7.5.10 *Glass*

Glass also originates from rock. It is manufactured from ground rock (soda-lime-silica). The most common type of glass is flat glass. This type is manufactured as clear float glass, patterned glass and wired glass. Tinted and other surface treatments are applied to float glass for special applications, such as solar controlled glazing.

Glass has very little thermal capacity but has a high resistance to moisture. Because of this, windows are particular targets of condensation. The glass is at virtually the same temperature as the outside air and the surface is an impermeable vapour barrier. Run-off from window condensation is a common cause of defects to window frames of both metal and timber.

Since June 1992 there have been statutory requirements in the use of glass extending the need to use safety glass for areas of risk. In particular these are glazed doors, low-level glazing, side panels of doors, and bathing screens. Many older properties have glazing which does not comply with these requirements. When re-glazing, the present standards should be applied wherever possible in the interests of safety.

Where glass forms part of a fire separating wall, e.g. between rooms, the glass is required to be wired, or of other special fire-resistance design. The size of panes is restricted so that the fire integrity of the wall is not compromised.

7.5.11 Metal

Metals are used for external protection (e.g. roofs and flashings), structural supports and for pipes, gutters, cables, etc. External protection is commonly provided by lead or zinc, and occasionally by copper. All of these materials will be found as coverings of roofs and as flashings at junctions between roofs and surrounding walls.

Metals have a high propensity to move with temperature variations. Correct installation of metals therefore requires allowance for such movement. The construction of movement joints, and the separation of large areas into bays which can move independently of each other is essential. Rigid joints, such as soldered joints in zinc, will have a tendency to fail or cause buckling and failure of the metal itself if there is insufficient allowance for thermal movement.

Structural metalwork, e.g. steel beams, are factory made and assembled on site. Joints are formed either by welding or by bolted or rivetted connections. For protection against both corrosion and fire, structural steel is often encased in concrete. Where the steel does not require corrosion protection, e.g. over an opening in an internal wall, fire protection can be achieved by plaster or other fire-resistant cladding.

7.5.12 Asphalt, felt and other bituminous materials

7.5.12.1 Asphalt

Asphalt is used as a covering to flat, or shallow pitched roofs. It is applied in heated form, when it is a liquid, and once cooled forms an impervious surface. It is finished with sand rubbed into the surface to minimise the surface being coated with a layer of its bitumen constituent, which will tend to craze. Solar reflective paints are often applied to asphalt to minimise temperature variations and oxidisation due to ultraviolet radiation. In recent years the extraordinary heat and duration of sunlight has caused exposed asphalt roofs to fail.

7.5.12.2 Felt

Roofing felt is a bituminous material, bound together with mineral or synthetic fibres to form a sheet. It comes in rolls and is laid in either cold or hot compound over the roof structure. Although today some roofing felts are designed as single layer coverings, the majority are laid in three layers, bedded in hot bitumen to form a single composite covering. As with asphalt, solar reflection assists in preserving the integrity of the roof covering and this is achieved either by chippings, a top layer of mineralised felt (i.e. small chippings embedded in felt) or solar reflective paint.

For both asphalt and felt, which are laid as impervious coverings, release of entrapped moisture or of vapour generated from below should be provided by pressure-release vents. These look rather like plastic mushrooms over a roof surface.

Both asphalt and felt are used for damp-proof courses and generally perform well. However, as both become brittle with age they are susceptible to fail where any structural movement occurs.

7.6 The enemies of healthy buildings

If the materials from which we constructed our buildings were inert, then the adverse effects of the elements would be limited. Buildings would be eroded by rain, blown down by wind, overloaded by snow, overheated by the sun. These phenomenon do take their toll on our buildings but, in general, building engineering has reduced these risks to a minimum.

In the UK we are fortunately unlikely to be faced with houses washed away by floods or in a state of near-collapse as a result of excessive winds. In the UK we are, however, faced daily with the more insidious degradation of buildings. Of all of the defects likely to be encountered, dampness is the most common and potentially the most harmful, both to the structure and the occupants

7.6.1 Water

Water can be an enemy of buildings in three ways: first, foundations can be affected by water in that it affects the structural stability of buildings by varying the supporting qualities of the sub-soil. When buildings are constructed, foundations are designed so that they will be unaffected by temperature or moisture content. Older houses, however, may have shallow foundations and be more susceptible to seasonal movement.

Second, and perhaps more commonly, it provokes unwelcome life. The source of life out in the garden is also the source of life within the dwelling. When water comes into contact with any organic material it will provoke life. It will also attract and sustain life.

Third, when inorganic materials such as brick, plaster or concrete become wet chemical changes occur. Carbon dioxide and sulphur dioxide are present in the air. When combined with water these can convert to harmful acids which will attack stone and concrete as well as brickwork.

The identification of the source(s) of the water is essential in order to specify a particular remedy, but whatever the source its effect is broadly the same. This is often forgotten, with statements such as 'It's not damp, it is only condensation' being promulgated all too frequently. (Appendix III helps the identification of sources of dampness.)

7.6.1.1 Sources of water

Moisture will pass through the structure by capillary action and direct flow. It will penetrate brickwork and concrete and will only be halted by an impervious barrier.

Damage from water does not only occur when there is a flood. Continuous and even intermittent, dampness will also cause defects and is, in the UK, the most common source of building defects.

DAMPNESS FROM THE GROUND

Rising damp   Even in a drought, the ground takes a long time to dry out. Even when extreme flood or drought conditions prevail, the soil on which we build our dwellings will not dry out. The water level and content in the soil may vary and, as seen elsewhere, this may cause structural problems, but the soil will still be damp and capable of transmitting that dampness to the building above.

Rising damp was the primary source of dampness in buildings in the last century. It was this defect which succeeding Public Health Acts and Building Acts sought to remedy along with insanitary drainage and washing facilities. The problem still exists more than a century later.

Rising damp can be confused with other sources of dampness. It has become the common name for all dampness originating out of the soil. However, its true meaning is only the moisture rising vertically within a wall or through a floor. This distinction is important when identifying the source of and liability for a defect.

In modern housing, impervious barriers to rising damp are unlikely to fail if correctly installed. Failure of older damp-proof courses such as slate, bitumen or asphalt is likely, due to the deterioration of the material coupled with structural movement which may fracture the membrane.

Remedial treatments such as silicone injection can be successful. However, the design and installation of these remedial treatments depends not just on the injection itself but a whole series of other factors. It is the failure to deal with these other factors which is the primary cause of recurrent rising damp.

A narrow definition of rising damp was given earlier. This is particularly relevant where remedial works have been carried out and where dampness recurs. In most cases, an injection damp-proof course will be covered by a guarantee. The guarantee rarely covers the recurrence of damp. What it warrants is that the damp-proof course injection will not fail. Therefore, when dampness recurs, but the injected barrier is still working, there will be no claim on the damp-proof installation.

It is with this in mind that other sources of dampness, associated with rising damp, but not actually rising through the structure, must be considered.

Lateral penetration   Lateral penetration does not only mean horizontal. Penetration of moisture through a wall or floor may occur where the damp-proofing is working but is being bypassed. Where external ground level is above, or close to the damp-proof course, moisture will pass across or around the impervious layer. Rooms below ground level are particularly vulnerable to lateral penetration. The external walls will be earth-retaining walls and in older properties will be protected externally with similar materials to those used for horizontal damp-proof course.

Modern basements are constructed with an impervious membrane on the exterior of the wall, linked through to the floor to provide a complete tank, rather like a swimming pool, but with the water on the outside. The floor and walls support the membrane and the load from the ground holds it in a sandwich between the structure and the soil.

In old basements, or any buildings where the floor is below external ground level, remedial works can only be successful if they replicate this modern construction. This does not require demolition and rebuilding, or excavation around the perimeter. It does require the sealing of the walls to the floor and the application of material to the walls which will withstand the hydrostatic pressure from the ground water. This remedial treatment, known as 'tanking' is common but nonetheless problematic.

Remedial treatments to older buildings are generally carried out by specialists. The level of expertise will, however, vary considerably, and the acceptance of responsibility for recurrence of dampness can be very difficult to achieve. Virtually all remedial treatments work well in the laboratory. The problems start when they are applied to real buildings by people.

The repair of dampness from the ground depends totally for its success on the condition of the base which is to be treated. Old walls often contain voids into which the injected silicone can flow, again not forming a complete horizontal barrier in the masonry. Broken brickwork to walls will not accept a silicone injection, neither will perished and friable bricks provide a stable base onto which cement tanking can be applied. Floors need to be stable and of adequate strength to accept a damp-proof membrane. If the base on which the repair is applied fails, then the repair itself will fail.

Bridging   Solid floors at or below ground level should contain a damp-proof membrane to resist the moisture from the ground. To provide a comprehensive barrier to dampness, this membrane must link to the damp-proof course treatment of the enclosing walls.

In cases of recurrent dampness in older properties, I have found that the most common failure is not with the injected damp-proof course, nor with the floor membrane, but is the link between the two. The reason for this is that the wall is treated by a specialist, the floor membrane is laid by largely unskilled labour as part of the placing of the concrete forming the floor structure. A little extra care at this stage would save thousands, if not millions of pounds across the UK.

Where the membrane has been covered by the concrete of the floor and not turned up at the wall abutments, the junction between the concrete and the wall will form a ready route for moisture.

Similarly, plaster internally or render externally which covers the damp-proof course will allow bridging, i.e. capillary action taking moisture past the membrane.

Salts and residual dampness   Moisture from the ground contains salts. It is the presence of these salts which assists in the identification of the source(s) of dampness (see Appendix III). When repairs are carried out, the salts will remain in the masonry. Unless the plaster contains an additive to act as a salt-inhibitor, these salts will migrate to the interior and cause staining and damage to decorations. The salts will also contaminate the plaster and absorb moisture from the atmosphere, resulting in recurrent dampness.

Treatment of a damp wall does not make it dry immediately. As a rough guide, the wall will dry out at the rate of 1 month for each 25mm of thickness. On this basis a 1-brick thick solid wall will take 9 months to dry out. In the absence of the need to use a building, it should be left unoccupied and gently dried out. Generally, however, this is not possible.

After a remedial damp-proof course has been inserted, the new plaster has to resist the migration of this residual dampness to the interior. This in itself will slow the drying process, as the moisture will then primarily dry externally. However, by applying the correct mix of plaster, some moisture discharge is possible internally without causing undue discomfort or damage. It is for this reason that, following damp-proofing work, only water-based, not vinyl, emulsion paint should be applied to the walls until the drying out is completed. Under no circumstances should impermeable wall coverings be applied for at least 12 months after treatment.

Dampness from the ground can therefore be due to:

a)   absence of an effective barrier to rising damp and/or
b)   absence of an effective link between the wall damp-proof course and the damp-proof membrane to the floor, and/or
c)   the bridging of the wall damp-proof course by render externally or plaster internally, and/or

d)   the migration of salts from the treated wall through an inappropriate plaster mix to the interior, and/or

e)   residual moisture in the structure.

ABOVE GROUND

Above ground in this context means above the damp-proofing layer. It therefore includes the lowest floor.

In buildings today, pipework is rarely run in a concrete floor. This is because the pipes are generally of copper which will corrode if unprotected in cement. The corrosion will perforate the copper, leading to leaks. Unfortunately the majority of buildings built up to about 1980 will have pipes buried in the floors. The practice has still not been eliminated, but it less prevalent.

Pipes leak. When they leak, water is discharged and causes damage. Weeping pipes, connections and fittings can cause substantial damage over a period of time, even though the leakage may be intermittent or at a low rate.

Plumbing installations are increasingly required to supply washing machines and dishwashers which are built-in or partly concealed by worktops, cupboards, etc. The concealment also hides the leaks and checking of these appliances may be necessary to find the failure.

From above   Often water entering at one point will not be seen directly below in the interior. Below the external covering is a whole series of layers of the onion, which will each absorb the water before it passes to the next layer down and eventually to the interior. Water will find the easiest route and even if it has penetrated directly to the upper surface of a ceiling, it will look for cracks and joints in the plaster to pass through to the room below.

A very small failure in the covering can give rise to significant water penetration. Once water has entered, it can have the same effects of vapour build-up and deterioration of the structure as water from any other source.

Cavity construction for walls   Cavity walls have a special requirement for damp-proofing. It is part of their design that water can enter the outer layer but will not penetrate to the interior. Around openings and wherever the cavity is bridged, by a floor beam for example, then a damp-proof membrane has to be installed.

As with so many parts of the building process, quality gives precedence to quantity and these damp-proofing measures were often omitted or installed inadequately. Remedial works can be expensive and disruptive.

Where insulation is installed in the cavity (see above) the choice of materials and the installation must ensure that a bridge is not provided to allow external moisture to penetrate to the inner surface. The materials used in new work are fixed so that an air gap remains between the outer face of the insulation and the inner face of the external skin. For remedial insulation systems, the use of moisture-resistant and largely impervious materials minimises the risk of water transmission.

FROM THE INSIDE

Condensation   Air contains water. It is held in vapour form. The hotter the air the more water vapour it can hold. The quantity of vapour in the air is measured as relative humidity.

There is no inherent problem with this moisture until the air meets a colder and impervious surface. The air is cooled and it can hold less water. The excess water condenses out onto the colder and impervious surface and forms water droplets. The water from this source has less salts in it than water from the ground, but the damage it can cause to buildings is not reduced.

The balance between heating, ventilation and insulation will need to be right to overcome the problem of condensation: it will be affected by moisture generation, but the basic failings of the building have to be present for any except the most extremely unreasonable usage to cause severe problems. The adjustment of any of these three factors will modify the conditions and may ameliorate the conditions, but it is only the balancing of all three that will cure the problem.

Relative humidity levels of around 50–60 per cent should not produce excess dampness and will maintain an adequate comfort level without excessive dryness. However, if humidity levels increase to 70 per cent or more then condensation will be commonplace and will sustain and generate the moulds and insect infestations referred to later.

Interstitial condensation Water vapour entering a flat roof structure struggles to escape or else condenses on the underside of a colder, impervious layer (usually the underside of the roof covering). This has two effects. First, it will cause the structure of the roof to become damp with all the consequences that follow. Second, the build-up of vapour in the roof will expose any weakness in adhesion of the roof covering, causing it to bubble upwards.

From construction Vast quantities of water are used in construction. It will be seen earlier in this chapter that most of the materials used in buildings either contain water or are mixed using water. All of this moisture has to dry out and be removed from the building.

Sulphate attack Brickwork is vulnerable to chemical attack. Sulphates from rainwater or from other parts of the building will react with Portland cement (a primary constituent of mortar). This reaction causes the expansion of the sulphate, in turn causing expansion of the brickwork. Where cavity brickwork is used, or where solid walls are constructed of different quality bricks for the inside and outside, this expansion can cause distortion of the wall.

Gases in flues can also often cause sulphate attack to the chimney, at which level the flue is not lined and the gases condense out on the inside face of the bricks.

7.6.2 Fungus

Fungal decay attacks all organic materials not just timber. Fungi are plants and grow and reproduce, their spores (seeds) are in the air and will germinate wherever there is an inviting host. In buildings we are concerned with those fungi which live off dead material. The identification of broad categories of fungal decay is necessary as the consequences of their action and the repairs required will vary widely.

7.6.2.1 Dry rot

Dry rot (*Serpula lacrymans*) is a fungus which has remarkable powers of survival. Thought to originate in the Himalayas it is perhaps a by-product of Britain's imperial past. As with other wood-rotting fungi it requires a moisture content of dead timber in excess of 19 per cent to start germination. What distinguishes it from other fungi, however, is its ability to survive and grow even when that moisture level is reduced. Dry rot will be able to survive in timber with a moisture content of 14 per cent. On the other hand where the moisture content is 40 per cent or more, it is unlikely to grow.

This ability to spread in relatively dry conditions, and to continue spreading after a water source is removed, makes it the most difficult to treat. It is able to travel through brickwork, finding passages in mortar joints and feeding off the hair and other organic material. It will attack furniture and belongings and can remain concealed until finally manifesting itself when too late.

Treatment will have to extend to at least 1 metre beyond the last growth and will include removal of all organic material. In older dwellings, this will include plaster which contains animal hair or other organic material as a binding agent.

When visible its first characteristic is warping or wrinkling of the wood, with often the outer, exposed, surface remaining intact whilst the material behind decays. The decaying material affected by dry rot has a cuboid cracking pattern, both along and across the grain of the timber. This cuboid pattern is also found with some wet rot fungi (see below), but in all cases erring on the side of caution is recommended, and where such patterns are found dry rot must be suspected.

In advanced stages, and only in concealed areas, mycelium in white/cream sheets with tentacles will be found on the surface of the affected material. When threatened the fungus seeks to reproduce and creates a fruiting body. These can be really beautiful and are usually flat, slightly puffy growths with white perimeters and rust red centres. The red part is the spore bank from which the spores are spread by the movement of air to new feeding grounds.

7.6.2.2 Wet rot

Wet rot fungi, of which there are thousands, if not millions, require high moisture levels to survive, usually in excess of 40 per cent. Once the water source has been cut off, the fungus dies.

The manifestations of fungal decay by wet rot fungi has a common thread, the distortion of the C and disintegration of the material. In most cases, there is also a dark staining which is a useful, but not 100 per cent certain, guide to distinguishing a wet rot attack from dry rot.

Remedial treatment is limited to the replacement of unsound material and chemical treatment of the affected areas. Chemical treatment must be carried out as although a wet rot attack can be cured by cutting off the water source, it is only too common to find that *Serpula lacrymans* takes over as soon as the moisture content declines.

7.6.2.3 Moulds

Many of the moulds found in dwellings are allergenic. Moulds require moisture (some of them more than 65 per cent moisture content), moderate temperatures (13–15°C) and suitable food. The food source is commonly wallpaper paste but may also be clothing and belongings. Any allergic reaction of the occupants will depend on the toxicity, buoyancy and concentration of the allergen. The most effective way to control mould spore allergens is by reducing moisture levels in the air. Increased ventilation can relieve these problems, must has to be balanced against temperature levels and the ability to maintain heat levels.

7.6.3 Insects

7.6.3.1 Wood-boring

These attack dead wood. Their presence is identified by their flight holes, usually small in diameter. It is the weakening of the timber by these holes and the internal passage cut out by the insects as food which cause the damage.

Detection in advance of some damage is difficult, if not impossible. The insects are minute, often not readily visible to the naked eye. The only manifestation of their presence is the flight hole after they have gone and a small pile of frass (digested and excreted wood dust) left behind, around and below the flight hole.

A tell-tale sign of death-watch beetle, which attacks oak, is the knocking sound it makes. It is unlikely that many housing managers will be involved in buildings using oak, and even less likely that they would be able to stay there in silence to hear the tapping sound. However, if buildings contain oak, an occupant describing tapping sounds in the middle of the night should not be dismissed as irrelevant to disrepair.

These insects have a seasonal life-cycle. The emerging insects lay eggs on the surface or within the wood. The eggs germinate and the grubs begin their lives within the timber. The grubs feed off the wood and after metamorphosis into flying beetles emerge by eating through the wood to the exterior.

Some, such as the death-watch beetle, do not emerge every year, but the presence of flight holes, especially if frass is visible, indicates that the infestation is active and that remedial treatment is required to prevent the subsequent generations from boring more holes and causing damage.

Treatment against common insects is now almost routine in rehabilitation work, and in new buildings pre-treated timber is normally used.

7.6.3.2 Disease carrying

COCKROACHES

Cockroaches are tough. They have developed a survival rate which has to be admired. They can remove themselves from one dwelling and contaminate another whilst the first is treated and then re-occupy it when safe to do so. For this reason, treatment of individual dwellings in a converted house or in a block of flats is unlikely to be successful. Whole block treatments are the only hope. Many local authorities have rolling programmes of whole block treatments which require repetition at frequent intervals to overcome this infestation.

Ants will also occupy homes, feeding off foodstuffs. They prefer damp, dirty and unhygienic food stores but as with cockroaches will require whole building treatment to eradicate.

Bedbugs, fleas (both human and animal) live in our homes. They can usually be treated effectively, but their removal will require the source to be treated also (e.g. the pet).

House mites are a major source of allergens. They are barely visible to the naked eye and are usually found in house dust and in bedding. The mite feeds on dead human skin. They require a high humidity level, not less than 45 per cent and therefore damp conditions increase their population and activity.

Silverfish, wood lice and other harmless insects invade the sanctity of our homes. These are sometimes inconvenient and unpleasant but do not carry disease or provoke allergic reactions.

7.6.4 Metals

Metals will corrode when affected by water but this is not just the effect of the water. Metals are affected by electrical action and this occurs when two different metals are linked by a conducting medium such as water. In buildings the most common metals which adversely react together are zinc and copper. Zinc is used not just in its own right, but as the galvanised coating to steel. Galvanised water pipes and water tanks joined to copper pipes will corrode.

7.6.5 Hazardous materials

7.6.5.1 Asbestos

Asbestos is in our buildings in many forms. Its use today is almost non-existent but it is the treatment of the material already present which has to be carefully considered.

The material is used in various forms and is itself of three distinct types (see Table 7.1).

ASBESTOS REMOVAL

Since as long ago as 1984 all asbestos removal contractors must be licensed by the Health and Safety Executive under the Asbestos (Licensing) Regulations 1983. The Asbestos Removal Contractors Association is a trade association attempting to maintain high standards. Contractors, including self-employed operatives, should be members of the Association and licensed by the Health and Safety Executive.

Table 7.1 Asbestos-based materials

Asbestos-Cement Boarding	Commonly used as roofing material, water tanks, flue pipes, etc. Contains about 10 to 15% chrysotile and 85 to 90% Portland cement, mixed with water. Flue pipes may contain amosite.	Wet stripping may be permissible if there is no prospect of breakage or fracturing, but where this is not guaranteed, then an enclosure should be provided.
Insulation Board	Commonly used as fire-resisting board. Contains up to 40% amosite or amosite mixed with chrysotile with 60 to 84% calcium silicate	Stripping can only be carried out using enclosures.
Rope and Gaskets	Commonly used to seal flexible joints. Can contain over 90% asbestos, usually chrysotile or crocidolite.	Stripping can only be carried out using enclosures

7.6.5.2 Glass fibre

Glass fibre wool is used in buildings as an insulator, usually in the form of matting. It will be found in lofts, where it is laid over the ceiling or under the roof and in cavity and timber walls, all to reduce heat loss. It is used for pipe and tank lagging and in a denser form as sound insulation quilt.

Glass fibre wool has raised similar concerns to asbestos in the past but is not considered to be anything like the same hazard. A recent report has concluded that it is not a significant health hazard. Nevertheless, strands will cause irritation to the skin and to breathing, and suitable precautions should be taken when handling.

7.6.5.3 Urea-formaldehyde foam

Urea-formaldehyde foam when mixed from its component parts will release formaldehyde. At low concentrations this will irritate the eyes and instil nausea and some breathing difficulties.

7.6.5.4 Radon gas

Radon gas is the result of the breakdown of uranium 238. The gas seeps out of the ground and will enter buildings. Increased insulation and restricted ventilation has resulted in a higher risk to occupants of dwellings. Concentrations vary throughout the UK but where high, increased incidences of lung cancer have been reported.

7.6.6 Users

Buildings would operate perfectly well if no-one occupied them. This may seem to be a ridiculous statement, but we cause no end of problems when we start to live in these structures. In particular we cause dampness.

7.6.6.1 Moisture generation

We all generate moisture by breathing, washing ourselves and our clothes, cooking and heating. For example, four people in a house for 12 hours will generate 2.5 kg of moisture. Cooking and food preparation can produce 3.7 kg per day, floor mopping 1.1 kg, clothes drying 12 kg and clothes washing 2 kg. The use of flueless gas heaters (e.g. calor gas) and paraffin heaters creates further water vapour. For example a paraffin heater will discharge 1 litre of water vapour into the air for each litre of oil burned.

Sweating/perspiration also generates moisture. Whilst we expect that sweating will occur at higher temperatures it must be remembered that this is related to humidity levels as well as temperature. For example at a humidity level of 22 per cent profuse sweating does not occur until 30°C but at a humidity of 60 per cent it will occur at 20–2°C.

It is the generation of these relatively large amounts of water by our normal use of the dwelling that the construction and design has to overcome by balancing insulation, ventilation and heating.

7.6.7 Animals

We also keep pets, which bring with them unwanted guests, and we produce refuse which we fail to properly dispose of or even to seal up pending disposal, and this encourages other animals into our homes.

7.6.7.1 Disease carrying

Rats and mice are probably the most common invaders. Rodents carry many diseases, mainly in the parasitic fleas which live on their bodies. They will, if cornered, attack humans and care must be taken in entering cellars or sewers where rats are likely to be present.

Pets, the great love of the British, are a primary source of pest importation into our homes. Fleas abound on dogs and cats and although they may not carry disease, having one's legs used as a food source is not entirely pleasant.

Disposal of faeces from pets is also a real problem, especially in blocks of flats. The effects of dog faeces on health have been well publicised and proper hygienic cleansing of areas used for defecating by pets must be maintained.

7.6.8 Plants, trees and bushes

It is also worth looking at the effect of trees and bushes on buildings.

In some instances, plant growth will be as a result of poor building maintenance. Where pointing to brickwork has been eroded, plants will often take root. It must be remembered that although we see plants as growing where we plant them in our gardens or window boxes, they have a life of their own. Seeds are carried in the air and will propagate wherever there is a source of food and water. They do not exclude parts of our buildings from this.

In other instances, domestic plants will expand beyond the confines of the plant pot and, particularly with creepers, seek out walls over which to climb. Plant covered walls can be attractive to look at but, with few exceptions, they will conceal the damage being caused to the fabric of the building. Plants need water and will obtain it from any source. The roots which are produced to search for and absorb moisture will be capable of finding a way through minimal openings, even between bricks. As these roots develop, they expand. The expansion will widen the gaps between the building elements, allowing more water to enter and setting in train a cycle of degradation.

Trees and bushes also have adverse effects on buildings, but primarily external. In hot summers buildings cracked. As will be seen below, the moisture in the soil, affected by drought or flood, will have a damaging effect on buildings. Trees and bushes need water to survive and in times of drought both the root growth, which will extend to search out water, and the extraction of the limited moisture from the soil will cause further damage.

7.6.9 Temperature and climate

7.6.9.1 Drought

The main effect is on the sub-soil and on the foundations, but high temperature levels also cause unusual thermal movement of the building elements, putting unexpected strain on the joints and junctions.

In times of drought, sub-soils such as clay shrink. The voids thus created allow for compression and slipping of the clay layers, which in turn removes support from the foundations of buildings. The subsequent cracking is known as settlement.

This drying out of the sub-soil will be hastened by trees and bushes extracting the remaining moisture. Often, therefore, serious settlement problems are a combination of changing sub-soil conditions coupled with tree growth.

7.6.9.2 *Frost*

The main effect is on the exposed surfaces of the building. Water expands when frozen. If water has entered the building fabric and then freezes, the expansion will cause damage, splitting bricks, breaking concrete. The opening thus created admits further moisture, continuing the cycle at an accelerated rate.

When tanks and pipes are not lagged, freezing will open joints and split pipes. When the temperature rises the water flows.

7.6.9.3 *Snow*

Snow imposes significantly increased loads on roofs. Modern roofs are designed for normal snow loads but in exceptional conditions structural failure is possible. Snow will also pull slates and tiles down roof slopes.

Rainwater pipes blocked with snow will not be able to carry away the water discharging from the roof when the thaw starts. There will be a time delay between the warming of the roof and the thawing of the ice in the pipe. During this period overflowing and leakage should be expected.

7.6.9.4 *Sun*

The sun discharges heat and radiation. The heat has a direct effect on buildings by causing expansion of materials, which in turn will cause fracturing of the less flexible elements.

Ultraviolet radiation from the sun will cause oxidisation of asphalt and break down its surface.

Temperature changes will affect some materials more than others

7.7 Taking the brief for the specification

The first operation is to assess the aim of the works. The interview with the client can seem like a cross-examination, but the specifier has to tease out as much detail as possible to minimise the perceived and real risks of variations for which no cost adjustment has been allowed. The aim must be to reduce the potential variables to the absolute minimum.

Without the aim of the works being codified it will be virtually impossible to monitor the development of the specification against the stated desire. Lists are the basic ingredient and assembling a list of desires which can be disassembled and analysed to describe the works required is a basic requirement. The list may start with a basic statement but will swiftly expand to develop detail. Supporting documentation will be required to further refine general performance requirements. These will include drawings and/or sketches to illustrate the extent of work; detailed diagrams to show how various components interact and are connected, and schedules of product types or even particular components.

7.8 Specifying damp-proofing repair

A damp-proof course (dpc) and other dampness repairs are common in refurbishment works and involve eradication of the causes, provision of barriers to dampness and replastering.

For guidance on assessing the nature and extent of dampness and on treatments, see Building Research Digest 245, which is the authority on dampness detection and repair. Such work is usually carried out by specialist sub-contractors who will issue a guarantee for the work they have carried out. Such guarantees are usually for 20 or 30 years. It is important to remember that it is the treatment, not the diagnosis which is guaranteed. The specification drawings should include details of the walling material, internal and external finishes, ground floor construction, and levels and positions of proposed horizontal and vertical damp-proof courses. This way the accurate specification and execution of the remedial work can be determined.

Internal finishes: the damp-proof course installer will normally require removal of any plaster/render along the line of the proposed damp-proof course to a height of 1 metre. It is therefore necessary to specify removal of salt-contaminated plaster and replastering. Replastering should comply with the damp-proof course installer's recommendations.

With any specialist damp-proofing treatment, in addition to the need to renew the wall plaster to a suitable specification the following works will be required.

1. Removal of all skirtings, architraves and other joinery to the affected walls.
2. Removal of radiators and any other fixtures or fittings within the work area.
3. Raking out of masonry joints and ensuring all traces of gypsum, dust or other friable material are removed
4. Use of only clean potable water and clean uncontaminated washed sharp/ rendering sand and fresh Portland cement is permitted.
5. An salt retarding additive approved by the installer must be mixed with the water before use.
6. For the replastering it is usual to apply a first scratch coat of 3 parts washed sharp/rendering sand to 1 part fresh ordinary Portland cement, incorporating water retardant.
7. Replacement joinery and skirtings must be treated with an approved preservative and primed on the back before being re-fixed.
8. Subsequent redecoration should be regarded as being temporary until the wall has dried out. Initial redecoration may consist of painting only with breathable paint (e.g. water based plain emulsion) after the new plaster has visibly dried out. Permanent redecoration should be delayed for at least six to nine months.

7.9 Specifying the formation of openings in walls

[Derived from http://local.diydoctor.org.uk and altered/updated]

This is one of the most common sections of works in domestic alterations. It is important to note that stud construction (studs are vertical members and horizontal stiffening members are 'noggins') can also be a load-bearing wall. The *only way* to be absolutely sure whether a wall is load bearing or not is to investigate along its entire length.

This type of wall is one that supports the structure or structures above it. The structure can be simply the floor, roof or the top half of the same wall. In other words, the wall may not just be supporting the weight of the wall above it, there may be floor joists fixed into the wall and in some cases, roofing beams and rafters. To provide an opening it is necessary to transfer the load above the new opening to the sections of wall either side. This will be carried out by use of a beam or lintel. The lintel itself takes the weight of whatever is being supported, so it important that this should be of a suitable specification for the circumstances and load. A structural engineer may be engaged to determine this.

The combined weight of these elements will need to be taken into consideration when working out the size of the lintel that is needed. The load above must be supported at all times.

7.9.1 Non-load-bearing walls

As noted above, some stud walls in old buildings may appear to be non-load-bearing but do carry loads. It is a common fallacy that hollow-sounding walls are only partitions and can be removed at will.

Similarly, constructed partitions may be non-load-bearing. These both comprise of vertical timbers either 100 mm × 50 mm or 75 mm × 50 mm. The thickness will vary due the height and the space available. Often the timbers include diagonal bracing, increasing the load-bearing capacity. Sometimes, in older buildings, the space between the timbers is filled with rubble (usually broken brick and cement) so the wall may sound more solid when tapped. This rubble infill adds slightly to the strength of the wall and increases its resistance to the passage of sound.

7.9.2 Doors

External and internal doors and frames can be bought in complete kit form and utilised as a prefabricated unit.

7.9.3 Plaster

Both sides of partitions are lined with plasterboard. Again thickness will depend on its function, thicker if increased fire separation or sound reduction is required

7.9.4 Floors and ceilings

In domestic buildings, upper floors in particular are often constructed of timber. Timber floors usually consist of the timber joists, a plaster ceiling beneath, floorboards on top and the nails or screws used to fix them to the joists. Together these are known as 'the dead load' and the floor or ceiling joists must be capable to support this load.

As an example, for a domestic property with timber floors and plasterboard ceilings, this dead load is generally taken by specifiers to be no more than 0.50 kilo Newtons per square metre (kN/sq.m is the measurement of the load). In the example of the bathroom that we noted earlier in the chapter, the weight placed upon a floor by way of the fittings and persons walking on the floor is known as the 'imposed load', and for normal household requirements, the imposed load will be about1.5kN/sq.m.

A source of guidance is the Building Regulations Approved Document A. Tables A1 and A2 list the size of joist necessary to support this weight, over a maximum span. Table A1 deals with timbers known as SC3 which have 'C16' markings from the supplier. These are for a dead load of more than 0.25 but not more than 0.50 kN/sq.m and allows for an imposed loading of no more than 1.5 kN/sq.m. Table A2 uses timbers known as 'SC4' which are high strength timbers containing very few, if any, knots. These timbers, are stamped 'C24' by the supplier. As these are not always commonly available, the more generally used timbers, dealt with in table A1 are used. These

7.10 Variations

There will always be variations during the course of building works. In an ideal world we would have no restrictions on possible options for the works, and at no cost variation, but this is not an ideal world. Whenever works, contract provisions or the details of components and fittings are varied there is a vulnerability to cost penalties. Even apparent cost savings by omitting work may not provide the full savings envisaged. But, let's not start with the downside. Building works must be approached with an air of guarded optimism.

The first draft of the works will be prepared by taking the brief for the specification.

Specification clauses are drafted by (a) establishing what you want you want to achieve (e.g. 'a nice bathroom') and then (b) by a series of questions and answers to establish more detail. This will require interrogation and investigation of the options available, bearing in mind the cost, availability and constructability. There is no benefit in specifying the best technical solution which might last for 100 years when the works are likely to remain for no more than 10 years.

A generous list of options should be assessed and reduced rapidly to a few. Availability of materials may be a determining factor as will be cost and complexity of use.

Having assembled an outline design and established some guiding principles, the specification falls into the following categories:

- Specification – setting out the standards sought for each section of work.
- Schedules – setting out the lists of specialised products, the recurring products (e.g. doors, windows, wall finishes).
- Drawings – to illustrate wall positions, pipe runs, outlet locations.

The combination of these items will provide a living and working document that will serve throughout the various stages of the building work and that will provide the basis for seeking tenders, codifying the contract, listing site activities, administering the contract and drafting of the final accounts.

The requirements for such documents are that when read with the drawings they provide an accurate description of the works for which prices are sought.

Specification language is partly jargon, but jargon with a purpose. The specification must be brief and to the point, it is not a novel or the script for an action film. The draft for the works must be based on good practice and tried-and-tested methods. Where any innovative works are to be carried out, full research must be evaluated.

Performance clauses should be drafted to establish a standard for the works, using recognised and available sources of technical information such as those referred to earlier. However, a level of cynicism should be maintained and works clauses revisited at each stage of drafting, as a single variation could impinge on other requirements.

Having taken an initial brief for the proposed works a first requirement is to research the materials and construction methods required for a successful contract outcome. For the lay client the sources of this information may well start with glossy magazines and TV programmes which entice us all to alter our homes and spend money on our property. The difficulty is how to transform an artistic and picturesque image into a physical reality. We all know the disappointment when the artist's impression of the toy is not realised when the box is opened and the 'all-singing, all dancing' image is revealed as a mundane, utilitarian item.

For the lay client, the manufacturers' documents from which a product is selected will probably be the catalogue. Whilst this provides a picture of the product, it may be idealised and not fully reflect changes borne out of development of the product as it is manufactured. To achieve a more accurate assessment of the product manufacturers' instructions/specifications are required with all their caveats and warnings. These will often provide useful information regarding fitting of components and details which would normally only be available after purchase. They will also provide details of limitations on use. Internet research used carefully and critically will provide user feedback and hints, or at least questions to be posed to the manufacturer.

7.11 Drafting the contract

The specification of the contractual provisions is an amalgamation of contractual provisions, material performance requirements and specific maters tailored to the particular project. Guidance is available with a number of standard specification guides and can be complex or simple.

'Preliminaries' are drafted as part of the specification to set out the variations and options within the standard form of contract which the parties wish to see. To illustrate the provisions required, we will use as an example the Joint Contracts Tribunal, Minor Building Works Contract.

The Joint Contracts Tribunal states that this form is appropriate where the work involved is simple in character; where the work is designed by or on behalf of the employer; where the employer is to provide drawings and/or a specification and or work schedules to define adequately the quantity and quality of the work; and where a contract administrator is to administer the conditions.

The form can be used by both private and local authority employers.

It is, however, not suitable where bills of quantities (detailed quantities to be re-measured as works proceed) are required; where provisions are required to govern work carried out by named specialists; where detailed control procedures are needed; where the contractor is to design discrete parts of the works, even though all the other criteria are met – consider the Minor Works Building Contract with contractor's design (MWD).

The contract documents can consist of drawings alone, drawings plus a specification, drawings plus schedules, drawings plus specification and schedules, or specification and/or schedules with no drawings. At tender stage the priced document must describe the works in sufficient detail to allow the tenderers to submit a price. The schedules of work incorporating preliminaries and reference specification will be utilized and must be comprehensive to enable the works to be priced without the provision of a bill of quantities. Simple quantities, for certain items or elements of work, may be included within the schedules of work. As the contract requires valuations to be carried out at monthly intervals, the pricing documents should be prepared to enable this to be easily done.

There are simple rules to guide the author of the conract. Words inserted into the contract to describe the proposed works should be succinct, brief but comprehensive and include the scope of the whole of the work.

For costing purposes there must be a 'base date' (see Chapter 8) to fix the date for calculation of the cost and to allow for a reference point to set the date from which the currency of certain documents is measured.

The guidance recommends that this date approximately ten days before the date for return of tenders.

The costs must include allowances for complying with CDM Regulations. The form of contract requires this to be defined by the need to clarify whether the project *is* or *is not* notifiable. A project is not notifiable under the CDM Regulations where it is not likely to involve more than 30 days, or 500 person days, of construction work, or it is being carried out for a residential occupier as a purely domestic project.

In order to ensure that the parties are fully aware of their liabilities, the contractor is required to make good any defects, shrinkage or other faults which appear within the rectification period. If nothing is inserted, the period will be three months.

For a contract of more than a few weeks, the contractor will require to receive progress payments, normally at monthly intervals for the stated percentage of the value of work properly executed and materials and goods stored on site. If nothing is inserted, the default percentage will be 95 per cent.

A payment of the percentage stated is to be certified 14 days after practical completion. If nothing is inserted, the percentage will be 97.5 per cent.

The contractor is required to send all necessary information to the architect/contract administrator before the end of the rectification period. Provided the certificate of making good has been issued, the architect/contract administrator must certify the amount remaining to the contractor, issuing the final certificate to the employer within 28 days of receiving sufficient documentation from the contractor. If nothing is inserted, the period will be three months.

Employers' liability insurance will enable the contractor to meet the cost of compensation for their employees' injuries or illness.

The Employers' Liability (Compulsory Insurance) Act requires that a statutory minimum level of employer's liability Insurance (currently £5 million) is maintained by the contractor.

Public liability insurance covers claims made by members of the public or other businesses; cover will vary depending on the type and location of work. The level of cover required should be inserted here.

These are alternative provisions depending on the nature of the parties and their willingness to accept obligations. It cannot be assumed that these provisions are accepted. For example, depending on the nature of the project and insurance available, the parties may use:

i) clause 5.4A on its own (where the works are not an extension to or an alteration of an existing structure);

ii) clause 5.4B on its own (where the works are an extension to or an alteration of an existing structure and the employer can obtain the insurance in joint names in compliance with clause 5.4B);

iii) clause 5.4C together with clause 5.4A (where the works are an extension to or an alteration of an existing structure and where the employer is a residential occupier and cannot obtain the insurance in joint names in compliance with clause 5.4B).

Under clause 5.4B, the employer is required to arrange the insurance. This would often be through the employer's existing buildings policy which, if the employer is a residential occupier, can be difficult to obtain. Problems may also arise when the property is leasehold and the freeholder is taking responsibility for the buildings insurance; contents insurance usually cannot be made to cover the joint names and getting the freeholder to extend their insurance may not be feasible.

The specifier/contract administrator must be cautious about giving insurance advice to the Employer and under no circumstances should they recommend a policy or fill in the forms for the Employer to sign. If in doubt, professional advice should be sought.

7.12 Specifying the details of the works

The specification is a key part of the tender documentation, which also includes the form of tender, instructions to tenderers, conditions of contract and questionnaires. The specification describes as accurately as possible the supplies, services or works that are required. In some cases it will be a highly prescriptive specification (for example, where a specific item of equipment is required). In other cases the specification may be output or outcome focused, where the regulations describe expected standards and results, but it is not necessarily prescriptive regarding how this is achieved.

Preparing a specification requires in-depth research to be carried out and a great deal of accuracy. A poor description of the requirements may mean that the product or service is not delivered to the standards required, and late changes to a specification may result in additional or abortive costs. A clear layout is required setting out the subject of the clause. For example, with a bath: its intended location (e.g. 1st floor rear room); the finished product sought (e.g. bath with a shower over); the preferred materials (e.g. white acrylic). This can then be extended to include the working sequence, for example install water serves and drainage, testing finalisation and acceptance.

On-line drafting tools provide an appropriate guide and drafts for works. The leading tool in the UK is the National Building Specification. However, as with all tools the secret of success is knowing how to use them.

Specifying the works requires an overview of the project as each operation will comprise of a series of work items in which several trades may be involved. Each of these trades comprises a separate work item and each of these trades may start and stop, re-start and return later during the construction process. It is not necessary to specify each stage in detail but the recognition of this intermittent interaction between trades is an essential component to the construction process and reflected in timetabling and costs. For example the formation of a new window opening in an existing wall might include:

- Needling through the existing construction to prove temporary support. This may require additional supports beyond the immediate work area (e.g. where support is required from the ground for works at 1st floor level).
- Demolition of the existing structure and then piecing-in and making good to exposed reveals.
- Cutting out for and installing a new lintel over the opening, including packing up to the existing construction and removing debris.
- Manufacture and installing of a lining to the new reveals.

This brief list does not include all the requirements for protecting against water ingress and provision of insulation.

7.13 Writing the specification

7.13.1 Pre-specification tasks

- Ensure that there is a clear understanding of what is required. If producing a specification for another party, identify and agree the customer or user's requirements and, again, ensure that they are fully understood.

- For works, analyse any existing provisions to determine the impact of the new proposals and the relationship of new to existing structures.
- Research the market by talking to suppliers, other purchasers, industry associations, etc., to identify possible solutions, indicative costs and delivery time-scales.
- Identify the possible risks associated with the procurement process so that ways of controlling the risks can be built into the specification.
- Identify the scope of the contract and what the range of goods and services are which the supplier will be asked to deliver.
- Identify the evaluation criteria so that the specification will reflect the importance of each criteria.
- Determine how you will monitor performance of the contract.

7.14 Structure of the specification

Specifications will vary in length and complexity depending on the nature of the product or service being procured. However, the following provisions are in most specifications, and should be included unless there is valid reason otherwise:

- Title page – Describe the project and identify the council and the main contact person (or contract administrator).
- Table of contents – Ensure the document is well set out and easy to read, using plain language.
- Definitions – In addition to the definitions in the contract part of the tender, it is important that there is a list of definitions, or a glossary of terms, to ensure that technical words and phrases in the specification are mutually understood. Failure to define key words and phrases may lead to misunderstandings and inappropriate solutions.
- Introduction – The introduction should briefly explain the requirement and the context of that requirement.
- Scope – The scope will address areas such as anticipated demand or volumes, whether the supplier is to supply only, supply and install, provide training, provide support documentation, etc. and, where appropriate, should identify specifically what is not to be included.
- Background – The more information a tenderer has, the better able he or she is able to respond to the tender. Background information may cover, where appropriate, the reason the council is tendering, its expectations, the implications for the council implementing the solutions, and other options which have been considered by the council (and if dismissed, why). It may also explain how the solutions may link into other requirements and applications whether already implemented or planned for the future.
- Service conditions and environmental factors – Explain any factors which may have a bearing on the operation of the goods or services. For example, if the physical environment may impact on the output design or performance, the specification must highlight these conditions. Examples are:
  - Operating and storage conditions.
  - The need for interchangeability or compatibility with existing services and equipment.
  - Personnel and health and safety aspects.
  - Existing facilities to be maintained throughout a contract period and what has to be done to ensure this.

The specification must also cover any particular sustainability requirements, for example energy usage and the recycling capability of the goods.

7.15 Evaluation criteria

The evaluation criteria that will be used to assess the submitted tenders must be developed at the same time as the specification. The evaluation criteria must reflect the key needs of users and should be included in the specification so tenderers can construct their bids accordingly.

Quality control

Whatever works are to be executed there will need to be quality control. At specification stage it is important to specify how this will be provide and by whom, as well as their power under the building contract. If these parameters are not set out at pre-contract stage it will not be possible to impose them at a later date without possibly prolonged and painful negotiation.

Monitoring of works on site is a role which is not implied into the contract. It is implied that he contractor will execute the works properly and in accordance with the specification, but the person monitoring this is the contractor

Building Regulation compliance (see Chapter 1) is achieved by intermittent checking by the inspector and in some cases works must be left open for approval. However, this generally does not apply to finishes but to issues affecting health and safety

We now look at how works are specified. These are adequately provided for in other publications and the authors do not consider it necessary to repeat detail here. Discussion of the contractual provisions follow in Chapter 8

7.16 Specifying a bathroom

Bathrooms are one of the most utilised rooms in a home, and getting these spaces renovated is often every homeowner's priority. This is because bathrooms have evolved from mere sanitary areas into refuges from the rigours of family and working life, and are increasingly well specified as a result. Besides more luxurious fittings and fixtures people are looking for more natural light in their bathrooms, as well as added convenience – with additional storage and double basins all there to make life easier. Planning is a major part of any bathroom renovation, so ensure both the client and the specifier take the time to consider all the various elements.

If you are planning a new bathroom, or undertaking an extensive renovation, the following elements need to be considered very carefully. Such a renovation will normally have to last several years, and therefore durability is a major consideration:

- Ceiling – will this withstand the steam generated by the shower and bathing as well as provide the necessary fire resistance and surface on which suitable decorations can be applied.
- Flooring – tiles should be chosen with aesthetics and practicality in mind, with a non-slip rating essential from a safety perspective. When stepping out of a shower this feature will be appreciated.
- Fittings – major fittings such as the toilet, basin, shower and bath should be carefully chosen, as they are likely to be with you for some time. Twin flush toilets make the most

of water saving features, showers may require a pump to obtain adequate pressure and with baths comfort is an important consideration.

- Taps – you get what you pay for with taps, with water saving features an essential characteristic as well as aesthetics.
- Lighting – adequate task and mood lighting is essential for any bathroom, and should ideally have individual adjustable controls, safely located.
- Storage –storage space should be maximised in every part of your bathroom, by installing new storage areas.

We now consider a few of these aspects in more detail.

7.16.1 Specifying concrete floors

The provision of accurate and user-friendly safety information by flooring manufacturers and suppliers to specifiers is essential to enable users of the flooring (employers) to comply with health and safety legislation in commercial settings. There is no similar statutory requirement for the homeowner but this is still a particularly important matter in relation to bathrooms where floors are more likely to be slippery. To prevent slips at work specifiers must consider a number of risk factors. These include the choice of flooring and how it is installed, used, cleaned and maintained, as well as contamination, footwear, behavioural and environmental factors. The use in domestic situations raises similar concerns

During the specification of new floors, too little attention is paid, in many cases, to how slippery the installed floor will be during use, especially in areas where floors may become contaminated with water, oil, grease or dust. This can result in the installation of flooring materials which quickly become slippery, leading to high numbers of accidents.

Limited information exists regarding the slipperiness of some commercially available flooring materials; information that does exist is often generated using inappropriate test methods. Slipperiness information tends to be aimed at specialists, which limits its practical use. It is therefore often difficult for specifiers to identify a floor with sufficient slip resistance.

Employers may be considered fortunate in that they have a duty under the Workplace (Health, Safety and Welfare) Regulations 1992 to make sure that their floors are suitable for their intended purpose. This duty concerns both the construction and the physical characteristics of the floor, including that it should not be slippery. 'Suitability' also has to take account of the circumstances of use, such as the likelihood of fluids or dusts contaminating the walking surface of the floor. The employer also has a duty to take reasonable steps to keep floors free from substances which could cause a slip. Such considerations should be incorporated into domestic situations wherever possible.

For new build and refurbishment projects in the commercial arena, clients and specifiers also have duties under the CDM Regulations (*The Construction (Design and Management) Regulations 2007*) – which refer to the Workplace Regulations – to consider the intended use of buildings, including floors. Installed flooring has to have sufficient slip resistance to enable the client to comply with any statutory duties. Manufacturers and suppliers of flooring must therefore take reasonable steps to ensure that users will not be exposed to risks due to the properties of their flooring.

The information provided by the supplier can be crucial to ensuring that the flooring is suitable for the intended use. Where a product is marketed or sold as being 'slip-resistant' or 'safety' flooring, information should be provided by the manufacturer to support these claims. Test data should be available and this must be presented clearly so that it is easily understood by the specifier.

Flooring suppliers provide a variety of technical data to describe the performance of their products.

7.16.2 Plaster

Remedial plaster works after dampness raises the following issues:

Masonry joints should be raked out and all surfaces thoroughly cleaned to remove dust or other friable material. Any organic matter (including timber fixings) must be removed. Cement-based plastering systems should not be applied to painted surfaces or those coated with bitumen. Any damp-proofing should be deferred for at least 14 days and any salt efflorescence removed before proceeding. The residual water in the wall must evaporate before normal dry conditions are achieved (one month per 25 mm thickness of wall is a rough measure).

7.16.3 Finishing: redecoration

After allowing the backing coat to set and dry (for a minimum of 24 hours, with the finishing plaster preferably applied at 1.5-3 mm thickness). As a finish, a high quality wood float finish can be used. All replacement skirtings, etc., should be either pre-treated or treated with a suitable preservative, and plugged and screwed with plastic plugs, or a suitable adhesive. Redecoration should be regarded as being temporary until the substrate has fully dried. Emulsion paint is recommended as a temporary measure.

8   Post-contract

'Post-contract' begins as soon as the specification is tendered and quotations are sought. The requirements for the basis of the contract are set out in Chapters 6 and 7. The moment the tender is sent out for quotations, the specifier is faced with the post-contract consequences of what has been sought.

The first set of queries arrives from the tenderers, and they will not all be the same. They usually arise as 'did you know that . . . no longer manufactures that'? This is why a last-minute review of the specification prior to tendering is an essential task. In some cases an alternative item which can perform the same function and with similar, if not the same, durability and quality will be available and can be specified as an option. In this way, all tenderers will quote on the same specification. Without ensuring all tenderers price for the same works, it is extremely difficult to evaluate the quotations received.

If post-tender/re-quotation variations occur, it is essential that supplementary information/changes are notified to all tenderers with adequate time to revise pricing. If these variations are minor, tenders can be sought and the variations negotiated with the lowest tenderers, usually the lowest two.

Having achieved an accurate contract, the drawings/schedules/specifications setting out the specified works will need to be assembled usually by the specifier to provide an unambiguous list of works and requirements on which all parties agree. The client should not seek to vary these works after tender without accepting that every variation will have a significant cost in both money and time.

With a finalised specification, works can commence. The client will want work to start on a Monday and be completed by the Friday, but this is unlikely. It is important that the tenderer is given time to assemble delivery information for materials and components prior to commencement. For example, a window which is on a delivery period of 3 months may delay the whole project if the entire works is due to last only 2 months. The contract period will need to be amended to reflect actual delivery times.

Therefore, in simple terms, a building contract is created when a house owner accepts the tender submitted by a building contractor to carry out construction works. The construction works may consist of the erection of a new building, or the renovation of an existing building. For the proper undertaking of the work by the building contractor, it is necessary to ensure the contract in writing. The use of a formal contract is advisable. The form of building contract is largely standardised, although the type varies, depending on which bodies are using them. In this book we have used the Form for Minor Building Works issued by the Joint Contracts Tribunal. (JCTMBW). Other contracts for civil engineering works are different from those used for building works.

For house renovation and even for the construction of a small building where costs do not exceed £200k, the JCT Form of Building Contract for Minor Works will be a suitable document to use. The contract is executed between the two parties, that is, the client and the contractor. The specifier is usually the administrator of the contract and acts in an impartial way in this professional duty.

It is also traditional for the specifier to be the administrator of the contract, acting as a quasi-arbitrator in administering and interpreting the meaning of the terms and conditions of the contract. However, in most contracts, an arbitration or an adjudication clause is also included; this provision enables any serious dispute between the owner and the contractor to be referred to the arbitration of a person to be agreed on by both parties. The adjudication provision usually does not apply to personal clients.

If a dispute arises, the contract document will be used as evidence to be submitted to the arbitrator or to the court.

8.1 What is covered by a building contract

The contract documents consist of the Articles of Agreement, the conditions of contract, and the specification and drawings, all of which are prepared by the specifier (called an architect or a contract administrator in the contract), who may be assisted by a quantity surveyor.

8.2 Scope of work

This describes the work proposed. It is a brief description of the intention of the person commissioning the works and briefly describes the contractor's obligations sought and codified in the form of contract. The detailed description of the works is usually given in the specification and drawings by the specifier. This section will probably appear as 'To refit the Bathroom at . . . .'

The contract, or rather the accumulation of documents which together form the contract, will include the following items:

- The programme.
- Procedure for dealing with variations or changes.
- Procedure for dealing with nominated subcontractors and suppliers (i.e. where the designer or person requiring the works to be supplied by particular manufacturers or carried out by particular trades people).
- The requirements for insurance by both the employer and the contractor.
- Payment terms and procedure.
- The maintenance period which by default will probably be 3 months.
- Grounds on which the contract can be terminated by either the employer and the contractor.

8.2.1 The programme

This is a time chart that the contractor is usually required to produce to show the sequence of work and the time to be taken to complete each part and also to include the relationship between different trades involved in the contract. The commencement (start) date and the completion date are agreed between the parties at the time of the contract. This is then called the contract period.

8.2.2 *Variations to works or design changes*

In repair and in renovation works, variations are inevitable once work gets going on site because it is rare that all hidden areas can be fully exposed for investigation before work starts. These variations therefore arise from unforeseen site conditions and circumstances not anticipated beforehand; for example, an uncooperative neighbour who will not allow access through his or her property, previously not expected to be a problem, or the fungal decay in hidden timbers becoming apparent. It is for reasons such as this that the specifier will specifier allow a contingency sum in the contract.

Variations due strictly to design changes initiated by either the owner or the specifier should be avoided or minimised as they can result in delay and extra costs.

8.2.3 *Subcontractor/suppliers*

The minor works contract empowers the specifier to nominated subcontractors to carry out work or supply goods or materials whenever a PC (prime cost) sum has been included in the specification. This provision gives the specifier, on behalf of the owner, the freedom of choice in the selection of a subcontractor or supplier for works which are of a specialised nature, for example, electrical subcontractor, air-conditioning specialist or interior decorator.

8.2.4 *Insurance*

The insurance clause is a standard provision and the contractor can be required to take up a contractors' All Risks Policy which even includes coverage for professional fees. Equally the client must ensure that any existing buildings or contents insurance adequately covers the proposed works.

8.2.5 *Guarantee/warranty*

This is usually only provided for works carried out by a specialist subcontractor or supplier of goods. One such example will be damp-proofing works or particular components such as bathroom fittings or widows. The specifier will normally recommend that the owner insist on these whenever specialist works are involved.

8.2.6 *Cost fluctuation*

In a JCTMBW contract, a cost fluctuation clause is not included as a condition. Such a contract should be completed in a short period of time, say within 4 months and material price fluctuations should therefore have no effect.

It is quite a common practice for the specifier to allow prime cost (PC) rates for particular materials, for example tiles, sanitary fixtures and specialist supplies. This PC sum sets the budget for the client and enables him/her to control the cost when selecting the item.

8.2.7 *Payment terms*

Interim payment is usually certified by the specifier at monthly intervals. However, in minor or renovation works the payment can be made at agreed stages of the works. There is usually a 5 per cent retention on any valuation of the works included in each payment. This works as follows: if the contractor's claim is £10,000 and the specifier agrees with the valuation, the client will only pay £9,500 plus any Value Added Tax at this interim stage. However, there is a limit to the retention, which is reduced to 2.5 per cent at the end of the contract. When the total contract value is small, the retention sum is sometimes increased to 10 per cent, which will be paid out to the contractor at the end of the maintenance period. The retention money is held by the client but belongs to the contractor. It must be kept safe and available to pay to the contractor when so certified by the specifier.

8.2.8 *Maintenance period*

This is a 6-month period or sometimes a 12-month period, after the completion date, during which time the contractor is liable to make good any defects in the works which are due to breach of contract or poor workmanship or the use of defective materials. The contractor will have to remedy the works at his own expense when the specifier notifies him in writing of any such deficiencies. Normally this is following an inspection by the specifier at the end of the maintenance period but for urgent matters may be referred to the contractor during the period. It is important that no other contractor is engaged to remedy any such defects as this could render the contractor not liable.

8.2.9 *Termination*

The conditions of contract provide both for termination of the contract on grounds of default by the contractor and for termination without default. The latter gives the client the right to terminate the contract but may be liable to the contractor for damages including loss of profit if any. The LCTMBW form sets out particular grounds limiting the basis on which a contract can be terminated The termination of a contract by either party often takes place under unpleasant circumstances and often leads to disputes and arbitration. Because of the specifier's role, he/she normally attempts to resolve such disputes.

8.3 Post-commencement

It is important to remember that the contract is between the contractor and the client. The specifier/contract administrator assists with contracts and acts as a coordinator between the contractor and client. An administrator's work duties may range from general office administration to technical document reviews and recommendation of solutions. It requires assessment and review of requests for information from the contractor and the client, possibly with separate agendas.

The contract administrator IS NOT a full-time quality control monitor. Where that function is required, then the client should appoint a quality control service. The *building regulation official* (either an approved inspector or a building control inspector from a local authority) will check compliance of what can be seen, but will not verify finishes against the specification or indeed performance of materials and works already covered by subsequent works. This building regulation role is to ensure compliance with building regulation and health and safety requirements only.

The appointment of a *quality control service* (often still known as the *clerk of works*) should be carried out prior to tendering and the specifier/contract administrator will need to draw together the roles of specifier, CDM co-ordinator, if applicable, and quality control. The terms of reference for quality control should be determined and contractually defined to include the frequency of visits and what particular operations on site require inspection and verification. The titles 'quality control' and 'clerk of works' are, for the purposes of domestic building works, often interchangeable.

The role of the clerk of works is based on his/her impartiality in ensuring value for money for the client, achieved through rigorous and detailed inspection of materials and workmanship throughout the building process. In many cases, the traditional title has been discarded and we now fine titles such as site inspector, specified building inspector and quality inspector, but the requirement of the role remains unchanged.

The clerk of works can be a very isolated profession on site. The inspector is the person that must ensure quality of both materials and workmanship and, to this end, must be absolutely impartial and independent when making these decisions and judgements. The inspector cannot normally, by virtue of the quality control role, be employed by the contractor – only by the client, reporting directly to the client (usually via the contract administrator on behalf of the client). The role is not to judge, but simply to report all occurrences that are relevant to the works.

In general terms, most clerk of works will have 'come through the tools', meaning they are trades people who move to the professional side of the construction industry often via further education.

The record keeping of whoever visits the site should be written and be accompanied by photographs. The following details should be recorded:

- Date and time of visit
- Persons present on site
- Weather conditions
- Material delivery and labour delays
- Information sought
- Commentary on works executed.

The quality control function is continuous and should be carried out so that all works which will be covered up by subsequent works are inspected before covering.

8.3.1 Building Control inspections

Building Control will also inspect the works, but this is no substitute for your own quality control. Building Control inspections vary depending on the type of work, and minor works usually need fewer inspections than more complicated projects. However, some inspection times are mandatory:

- Commencement of work
- Excavation of foundations before concreting
- Foundations when constructed
- Damp proof course when laid
- Site concrete or floor slab (before being laid)
- Drains (before backfilling)
- Drains (after backfilling)
- Prior to occupation
- Completion of work.

Most of the above elements relate to works that may be covered up but are not necessarily the only ones the inspector will make. Some projects require additional inspections specific to the work, such as reinforcement of concrete foundations and floor slabs, fire separation within the structure. The inspector needs to inspect prior to covering up (e.g. before plastering, to enable the work to be inspected before being covered up). Failure to give notice of when mandatory inspections are required may lead to having the work uncovered for inspection. Ensuring that the council/inspector is notified of the various stages can avoid conflict between the owner/builder and the inspector.

When the work is completed the inspector should be notified, so that he or she can carry out a completion inspection. Providing the work is in accordance with the appropriate regulations, a completion certificate is then issued. This is important as proof that the project complies with Building Regulations and it may be requested if the property is sold or the client wishes to raise a loan secured on the property.

Not all aspects of building work are covered by the Building Control inspections, for example:

- Finishing and decoration
- Standard of workmanship.

The Building Control completion inspection will include a collection of data relating to fittings and services, such as test certificates (e.g. electrical), certificates of acceptability (e.g. FENSA certificates for glazing, thermal insulation calculations, etc.).

The owner of a building or the person having the work carried out is legally responsible for complying with Building Regulations. This includes applying for approval to carry out the work and ensuring the work on site meets with the regulations. The builder is only responsible for notification of the statutory inspections. If there isn't a main contractor then the client is also responsible for the notices.

If a notification is missed, and where works have been carried out without all necessary inspections, a regularisation application is required. Completed or concealed work may then have to be exposed, e.g. foundations, structural items or drainage. These can then be inspected along with other aspects of construction to ensure they comply with the regulations that were in force when the work was carried out. The risk here is that remedial works may be required.

8.4 Post-contract issues

The tender quotation which forms the contact is a fixed price for carrying out the work, which cannot be changed once accepted by the client. This is expressed in the priced specifications that form the basis of the contract. Comparing quotes from several builders is not straightforward, as pricing is not the only variable. Dependability, willingness to rectify faults, attendance on site, prediction of events, etc., all distinguish a good contractor. The ambition to get the building project completed in the shortest possible time frame and on budget with no variations is common. But is it achievable? Usually at a price, and especially if you hire a larger contractor with a well-organised workforce and good experienced project managers. For this you will need a set of accurate drawings and specification documents with every detail finalised. This level of organisation should allow your project to progress efficiently.

At the other extreme in the building market is the small contractor, who may have at most two or three employees on his or her books, with the bulk of the building work undertaken by a network of subcontractors that he or she can pull in, as and when needed. They, too, will be able to provide you with a quote for your work, and will agree to work to a form of contract. With smaller overheads to maintain, and a less expensive workforce, the smaller contractor's price is likely to be considerably lower than a large contractor's. The quote you receive, however, is unlikely to be anywhere near as detailed as the one provided by the main contractor, making the competing quotes difficult to compare. In theory, whichever route you choose, you should end up with the same house.

8.4.1 *Access equipment/scaffolding*

[Derived from http://business.highbeam.com/410604/article-1G1-138187663/duty-surveyors.]

Before building work proceeds, scaffolding may be required, which in itself raises issues of access and ability to inspect. This can affect the public and also the neighbours. The following case is a warning for all involved, particularly for specifiers and quality control professionals. This case arose under the law of tort. A 'tort' is a breach of a non-contractual duty for which the law provides a remedy. There are a number of categories of torts, the broadest of which is negligence; another example is the nuisance factor, which allows an aggrieved party the right to seek a remedy for acts that interfere with the use and enjoyment of land. As an example of what can go wrong we can look at the case of *Mistry v. Thakor and Roberts*:

In this case the Court of Appeal had to re-examine a decision concerning an action brought in the tort of nuisance.

In July 2000, Mr Mistry was walking along Belgrave Gate, a street in the centre of Leicester. As he passed in front of a building owned by Mr and Mrs Thakor, two pieces of concrete cladding fell from the building, struck him and caused him serious injury. A claim was brought on Mistry's behalf in nuisance. In defending the action against them, the Thakors brought in as third party defendant, Mr Roberts, a chartered surveyor. The Thakors owned a number of properties and employed Mr Roberts as their property manager. Mr Roberts' terms of engagement included a duty to inspect the Thakor's buildings at least twice a year. The Thakors contended that Roberts was in breach of this duty, without which the injuries to Mr Mistry would have been avoided.

The building was built in the 1960s. On the front elevation along Belgrave Gate at both first and second floor level, there were courses of concrete facing panels each roughly 100×50 centimetres in size and weighing about 50 kilos. The panels were fixed to the wall by metal fixings and seated upon a steel angle which acted in effect as a shelf. In court, the experts agreed that the panels had fallen due to corrosion of the fixings behind the panel. The effect of the corrosion had been to push the bottom edges of the cladding panels away from the building. One of the experts had commented that 'the movement would obviously sever any bond with the original cement mortar bedding and the slabs that fell must have slipped off the corroded edge of the base angle'. It was not, therefore, in dispute that the building was unsafe and a public nuisance up to the point when the panels fell into the street causing the accident.

The case against the Thakors in nuisance depended upon whether they could be presumed to have known about the state of the building. The case against Roberts was that he had failed to identify the dangerous state of the panels, failed to have the dangerous state rectified and failed to inform the Thakors of the dangerous state and the steps which needed to be taken. In their defence, the Thakors claimed that Roberts did not know of the relevant defects to the panels and accordingly no knowledge of those defects could be imputed to them. Their duty as building owners had been discharged by instructing a reputable chartered surveyor to manage the property for them. The judge at first instance rejected this defence. Roberts must have known about the potential failure of the panels by a simple visual inspection from the ground floor. The fact that he attached no importance to the obvious defective state of the panels did not alter the fact that he had sufficient knowledge of the likelihood of a public nuisance. For the purposes of the law that knowledge would be imputed to the building owners, the Thakors.

An interesting aside in all of this is that the movement of the defective panels was clearly visible from street level. This plainly merited closer inspection, when the corroded fixings would have been evident and the need to take remedial measures obvious. Unfortunately, Mr Roberts had been unwilling to climb scaffolding. Not using the scaffolding which had been erected on the face of the building was described by one of the experts as bizarre. Roberts had recommended to the Thakors that in order to comply with health and safety legislation a building contractor should be engaged to look at the panels. It became clear during the trial that when using the expression 'health and safety' Roberts was not concerned with the health and safety of the public, but his own health and safety.

The judge was highly critical of Roberts for his refusal to climb scaffolding, which he described as extremely unusual conduct on the part of a chartered surveyor. He proceeded to find the Thakors liable for the claimed damages and in the third party proceedings held Roberts to be responsible for a contribution of 80 per cent of those damages. The Court of Appeal refused to interfere with the discretion exercised by the judge to award damages on that basis. Lord Justice Pill noted that professional men are employed to deal with things normally expected of them in their profession. Mr and Mrs Thakor did not expect that they would have to go to someone else to do the comparatively elementary task of climbing scaffolding to inspect their building.

From the above it can be seen that (a) liability can accrue to persons who are not party to the building contract (the public) and (b) that a professional should carry out a full inspection. The absence of inspection can also give rise to a claim for extra works on areas where the contractor and/or the specifier allegedly could not inspect.

It is therefore imperative that careful observation and recording of risks is made. The mere avoidance of the inspection is no defence to any future claim.

8.4.2 *Additional costs for the same work*

Extra work arising from deficiencies in the specification or lack of forethought will be the liability of the client, although the client will no doubt seek to place the liability with the specifier. For this reason detailed records of investigations and decisions take during the pre-contract specification stage are essential and agreement on the sharing of risk for matters which cannot be fully certain until works commence on site needs to be made. For example, a specified component which is modified by the supplier after contract or is no longer available in the same form is a risk which is neither the liability of either the client or the specifier. In those circumstances, the solution lies with the client working with the specifier. Of course, had the specifier prior knowledge of non-availability and did not forewarn the client, the liability is likely to fall to the specifier

8.4.3 *Extension of time and non-completion*

Under the JCTMBW building contract, matters not within the control of the contractor will accrue an 'extension of time' and any costs incurred will be payable to the contractor.

If the works are not completed by the contract completion date, or any extended date as agreed by the specifier or the parties, then damages may be deducted by the client.

8.4.4 Base date

The base date is a 'line in the sand', a point at which any notifiable additional costs to the contractor, e.g. labour rates, are fixed. It is deemed that all such costs are calculated at the base date.

When there is bad weather, the contractor may claim an extension of time. Bad weather is not merely cold or heat or even rain. It must be exceptionally bad and must affect the execution of the work. The severe winter conditions in the UK at the end of 2010 are a recent example. If a contractor is to seek to claim for delays encountered due to weather conditions, detailed notes of the weather encountered will be required, together with details of why alternative tasks could not have been executed. When it becomes apparent to the contractor that the progress of the works is likely to be delayed, the contractor 'shall forthwith give notice to the specifier' of the circumstances, including the cause of the delay. In other words, matters cannot be allowed to drift.

There is no definition of what 'exceptionally adverse (bad) weather conditions' means and while any weather conditions may cause a delay, only *exceptionally* adverse weather will give rise to any entitlement to claim reasons for delay. Whether or not the weather is sufficiently adverse will be a matter for the discretion of the contractor. It is therefore important when making a claim that the contractor includes sufficient evidence to demonstrate that the conditions are 'exceptionally adverse'. This might be done with reference to past records from a local weather station. Whatever the circumstances, record keeping is crucial.

8.4.5 Lack of labour

With small projects this is less likely to be relevant unless highly specialised skills are required. However, if a medium or larger contractor is engaged on small works, there is a risk that larger, more prestigious projects may rob smaller projects of labour.

Lack of effort and non-attendance usually arises when the contractor has too many contracts. Where one project drags on, or is delayed, it will have a 'knock-on' effect on other parts of the project. When this is caused by a temporary hiccup it may be possible to accommodate it but too often this is a sign of further delays caused by execution of works out of sequence

8.4.6 Disputes

The greatest area of dispute is, as one would expect, about costs and in particular variations. Variations do not always accrue a cost addition. They can reduce or increase the scope of works, but not necessarily the costs. How can the liability for such delays be assessed? To some extent the following may assist.

8.4.6.1 Variations caused by changes of mind by client

Even assuming that there has been no execution of works, the probability is that some commitment to a particular type of construction will have an effect on other works areas. For example, changing from a wall constructed in brickwork, albeit non-load-bearing, to construction using a timber frame will affect the fixings, the trades used to construct the work and the finishes available. Each choice of variable will also raise questions as to cost, speed and availability.

This assumes that the variation is a clear choice. However, the variation can also involve unforeseen obstacles, e.g. the discovery of a subterranean void, possibly an old excavation which requires a redesigned floor slab to overcome.

8.4.6.2 *Variations due to statutory requirements.*

These requirements may vary. Although there are provisions for changes, these do not always fully accommodate the changes required. For example if a window repair develops into a replacement, with the consequent upgrading of installation and glazing requirements, the trigger for the substantial variation of the upgrading of glazing/insulation is the change from 'repair' to replacement of a new window, which may appear as a minor change but moves the works into a different area of regulation.

Just to complicate these matters further, some contractors play 'hunt the extra'. In other words, having found a (possible) trigger, the contractor will seek to maximise the cost and time impact on the project. Again using the window example, the additional time and cost to design a new window and then submit an application can all be added to the costs.

During the course of a contract, *interim payments* are payable at pre-agreed intervals, but always in arrears. It is imperative that payments are for work carried out, and not 'on account' payments for works yet to be executed.

Fixed costs are exactly as stated. An example would be a statutory fee payable at a pre-agreed rate, e.g. statutory fees for the local authority building control inspections.

In the contract, approximate costs will require reassessment as they become more refined and defined. These approximate costs need to be adjusted in the course of future payments when they will need to be fully valued against actual costs.

8.4.7 *Daywork*

[Derived from http://www.whichbuildingcontract.co.uk altered and adapted]

In general terms, daywork is best avoided. It can prove costly to the client and is largely un-tendered. A good source of information is 'Definition of Prime Cost of Daywork Carried Out Under a Building Contract', based on contributions by the Royal Institution of Chartered Surveyors, the Building Cost Information Service and the Construction Confederation. It was published in June 2007, and superseded the previous edition from 1 September 2007. There are some significant differences between this edition and its predecessor. The new definition offers two options for dealing with the prime cost of labour:

- Option 'A' – 'Percentage addition' is based upon the traditional method of pricing labour in daywork, and allows for a percentage addition to be made, to the prime cost of labour applicable at the time the daywork is carried out for incidental costs, overheads and profit.
- Option 'B' – 'All inclusive rates' includes not only the prime cost of labour but also includes an allowance for incidental costs, overheads and profit. The all-inclusive rates are deemed to be fixed for the period of the contract. However, where a fluctuating price contract is used, or where the rates in the contract are to be index-linked, the all-inclusive rates shall be adjusted by a suitable index in accordance with the contract conditions.

Whoever puts the contract documents together (the specifier) can decide which of the above methods is the most appropriate. Consideration should be given, for example, to length of contract, whether the contract is fixed, to firm or fluctuating prices, or whether the costs are to be index-linked. Using Option 'B' gives the client price certainty in terms of the labour rate to be used in any daywork during the contract, but there is the potential that the rate will be higher, as the contractor is likely to build in a contingency to cover any unknown increases in labour rates that may occur during the contract period.

For larger and longer-term contracts, if Option 'B' is used and the rates in the contract are to be indexed-linked, BCIS publish a number of cost indices that may be suitable for such the circumstances.

8.5 Valuing works costs and the payment process

In order to assess works costs and the amount to be paid to a contractor, full details are required of any claim for interim valuations and, more importantly, for the final account. As a general rule the more detail supplied, the swifter and more successful the application for payment. Cash flow is obviously an important consideration and the minimisation of delays is essential for all parties. The application for payment should be submitted on the basis of work carried out. Depending on the contract provisions the value may include materials not yet fixed on site. The contract administrator must be able to assess the application and verify the proper execution of the works for which payment is claimed.

The process usually adopted is that all such claims are submitted to the contract administrator for assessment and, following acceptance, the issue of a certificate for payment. The contractor then submits an invoice to the client based on that certificate.

8.5.1 Loss and expense payments

This is a payment due to the contractor for any loss of profit on works omitted or possibly any loss of opportunity costs arising out of delays for which he/she has no liability.

8.5.1.1 Liquidated or ascertained damages

These are assessed on a pre-determined scale at a rate set out in the contact. The rate cannot be punitive and must reflect the 'tone' of the contract. The cost must bear a relationship to actual losses suffered or likely to be suffered. A scale removed from reality will be viewed as a penalty clause and may not be enforceable.

Actual damages payable by the contractor for delayed or non-completion will be based on codified losses or costs incurred by the employer as set out in the contract documents in a clearly intelligible formula.

8.5.2 Settlement of disputes

In order for disputes to be resolved, the parties will need to assemble and verify their facts to enable the following to be readily visible to all:

- Demonstrable accuracy
- Calm approach, time to reflect
- Legal intervention if required.

8.5.3 Dealing with disputes

[Derived from http://consensusmediation.co.uk/mediationnews.html]

Options available if the parties cannot resolve differences can include going to court or referring the matter to arbitration or mediation. Arbitration is where an 'expert' reviews the case and reaches a judgement and mediation is a route where the parties in a dispute are encouraged and helped to find a solution which they can each accept. A third route is 'adjudication', but this does not apply to owner/occupiers. Adjudication is a rapid method wherein under the contract an expert will decide upon a resolution, which is usually binding on the parties.

To assist the choice of route, case law is useful and attention is drawn to the following, with particular emphasis on 'alternative dispute resolution' (ADR):

Halsey v. Milton Keynes NHS Trust (May 2004)
The decision in this case established three new principles, which lawyers should note:

1. 'The value and importance of ADR have been established within a remarkably short time. All members of the legal profession who conduct litigation should now routinely consider with their clients whether their disputes are suitable for ADR.'
2. 'The fundamental principle is that a departure [from the general rule that costs follow the event] is not justified unless it is shown (the burden being on the unsuccessful party) that the successful party acted unreasonably in refusing to agree to ADR.'
3. 'The fact that a party unreasonably believes that his case is watertight is no justification for refusing mediation. But the fact that a party reasonably believes he has a watertight case may well be a sufficient justification for a refusal to mediate.'

The *Halsey* case may have been misunderstood, particularly by parties who believed that their case(s) were watertight. This issue has been addressed by the Court of Appeal in the case of *Burchell v. Bullard (2005 EWCA 358)*. The case concerned a building dispute where the claim was for £18,300 and the counterclaim was for £100,000. In the event, the claimant was awarded his claim in full and the counterclaim succeeded only in the sum of £14,300. The costs of both parties amounted to over £185,000 – a figure which Ward LJ described as '*horrific*'.

Ward LJ said in his judgment, 'The defendants behaved unreasonably in believing, if they did, that their case was so watertight that they need not engage in attempts to settle . . . The stated reason for refusing mediation, that the matter was too complex for mediation, is plain nonsense.' He went on to say, 'Halsey has made plain not only the high rate of a successful outcome being achieved by mediation but also its established importance as a track to a just result running parallel with that of the court system.'

Litigation lawyers should take note, therefore, that the preference of the judiciary towards 'mediation before litigation' continues. The risk of a successful litigant not obtaining his costs at trial if he refuses mediation overtures increases with each new step in the Court of Appeal.

It is important that any disputes are resolved effectively. Set out below are a number of other important ADR/Costs sanctions decision cases.

In the Court of Appeal case *McMillan Williams v. Range* (2004), a solicitor who received advance salary in excess of her actual earnings had to repay the excess to her employers when she left the firm. The court at first instance advised mediation, but both parties refused. The court ordered both parties to bear their own costs.

In another Court of Appeal case, *Dunnett v. Railtrack* (2002), an award of costs was denied to a successful party because it had earlier flatly refused to mediate. This case was a benchmark towards the requirement to mediate and followed earlier cases warning of likely costs sanctions (see *Cowl v. Plymouth City Council* below).

In *Hurst v. Leeming* (2002) it was decided that in appropriate cases it may be acceptable to refuse to mediate, but it is *'a high risk strategy'*. The critical factor in coming to a decision on the reasonableness to mediate is whether the mediation had any real prospect of success. A refusal will only be reasonable if, objectively, that prospect does not exist. This decision has, however been modified by Halsey (see above) in that whilst Hurst placed the burden on the successful party who refused mediation to justify that refusal, Halsey places the burden on the unsuccessful party to show that mediation had a reasonable prospect of success.

In *SITA v. Watson Wyatt and Maxwell Batley* (2002) a successful litigation party refusing mediation escaped costs sanctions because the invitation to mediate was made at short notice in an effort to 'dragoon, browbeat and bully' and in a way that was 'disagreeable and off-putting'. The mediation proposal was a litigation tactic rather than genuinely designed to seek settlement.

In *Cable & Wireless v. IBM United Kingdom Ltd* (2002) a mediation clause was enforced by the court adjourning the litigation. The clause was not a mere agreement to negotiate and therefore unenforceable, but a real contractual commitment to find solutions 'which are mutually commercially acceptable at the time of the mediation'. Mediation is described as 'a firmly established, significant and growing facet of English procedure'.

In the Court of Appeal case *Cowl v. Plymouth City Council* (2001) the Court commented:

> Without the need for the vast costs which must have been incurred in this case . . . the parties should have been able to come to a sensible conclusion as to how to dispose of the issues which divided them. If they could not do this without help, then an independent mediator should have been recruited to assist. That would have been a far cheaper course to adopt. Today, sufficient should be known about ADR to make the failure to adopt it, in particular where public money is involved, indefensible.

It is clear that who fail to consider ADR will be criticised by the court. The cases of Hurst and Dunnett and Halsey make it clear that any solicitors who fail to give due consideration to ADR may well find themselves at the wrong end of a negligence action by a winning client who fails to recover costs.

In addition the following practice advice was issued jointly on 22 April 2005 by the Law Society's Civil Litigation Committee and its Alternative Dispute Resolution Committee:

> This practice advice relates to the giving of information on mediation and other dispute resolution (ADR) options to clients before, and during the process of resolving any disputes between the client and third parties. The principle of why this advice and information should be given is to be found in the *dicta* of Lord Justice Dyson in the case of *Halsey v. Milton Keynes NHS Trust* and *Steel v. Joy* [2004] EWCA 576: 'All members of the legal profession should now routinely consider with their clients whether their disputes are suitable for ADR'.
>
> Solicitors should note that the court has a duty to encourage parties to co-operate with each other in the conduct of the proceedings – Civil Procedure Rules 1998 (CPR), rule 1.4(a) – and to likewise encourage parties to use mediation or some other alternative dispute resolution technique in appropriate cases – CPR rule 1.4(d). Where the

parties cannot agree to use mediation or another ADR process, the obligation is on the party wishing to use mediation or another process to say why it is appropriate in the circumstances. Section 2 of the guidance summarises the factors to consider in ascertaining whether a case is suitable for ADR.

The term ADR means both mediation and any other alternative to formal litigation or arbitration that might be an appropriate alternative means of resolving the dispute in the particular circumstances of the case. This might include expert evaluation, early neutral evaluation or conciliation, as well as mediation.

This practice advice applies to advice and information at the appropriate time, which may be at the commencement of a dispute within the initial advice, or at any later stage of the dispute. Practitioners should keep these options under review throughout the course of the matter. Solicitors should:

In appropriate cases, and at appropriate times, explain to clients whether there are ADR techniques that might be used other than litigation, arbitration or other formal processes; what those alternative processes involve, and whether they are suitable in the circumstances; and

Keep the suitability of mediation and other ADR techniques under review during the case and advise clients accordingly.

In assessing whether a case is suitable for mediation or some other form of ADR, the solicitor should have in mind:

- The nature of the dispute;
- The merits of the case;
- The extent to which other settlement methods have been attempted;
- Whether the costs of the ADR process would be disproportionately high;
- Whether any delay in setting up and attending the ADR process would have been prejudicial to the client; and
- Whether the ADR process had a reasonable prospect of success.

Solicitors should be aware that failure to provide information and advice at the appropriate stage may have costs or other consequences.

In summary, do not look to resolve the dispute but rather work to prevent the dispute arising.

Index